NEVER PANIC EARLY

NEVER PANIC EARLY

AN APOLLO 13 ASTRONAUT'S JOURNEY

FRED HAISE
with BILL MOORE

Smithsonian Books
WASHINGTON, DC

Published by Smithsonian Books
Director: Carolyn Gleason
Senior Editor: Jaime Schwender
Assistant Editor: Julie Huggins

Edited by Karen D. Taylor
Designed by Gary Tooth / Empire Design Studio

This book may be purchased for educational, business, or sales promotional use. For information, please write: Special Markets Department, Smithsonian Books, P.O. Box 37012, MRC 513, Washington, DC 20013

Library of Congress Cataloging-in-Publication Data
Names: Haise, Fred, 1933- author. | Moore, Bill, author.
Title: Never panic early : an Apollo 13 astronaut's journey / Fred Haise
    with Bill Moore.
Identifiers: LCCN 2021053466 (print) | LCCN 2021053467 (ebook) |
    ISBN 9781588347138 (hardcover) | ISBN 9781588347145 (ebook)
Subjects: LCSH: Haise, Fred, 1933- | United States. National Aeronautics and
    Space Administration--Biography. | Apollo 13 (Spacecraft) | Astronauts--
    United States--Biography. | Space flight--History.
Classification: LCC TL789.85.H35 A3 2022 (print) | LCC TL789.85.H35
    (ebook) | DDC 629.450092 [B]--dc23/eng/20211220
LC record available at https://lccn.loc.gov/2021053466
LC ebook record available at https://lccn.loc.gov/2021053467

Printed in the United States
26 25 24 23 22   1 2 3 4 5

*Dedicated to the more than 400,000 participants
in the Apollo program who made what seemed
to be an impossible goal attainable.*

TABLE OF CONTENTS

FOREWORD
BY GENE KRANZ

Hundreds of books have been written by astronauts, and while reading Fred Haise's early, well-written chapters, I concluded that *Never Panic Early* serves two purposes. First, it's the story of the Apollo generation of astronauts. Second, it recounts Haise's determination and destiny to become a member of that select group.

I don't remember the first time I met Haise, but we became very close in the post-Apollo years. We are forever brothers in a fraternity of those who have taken flight, and, because we fly, we envy no man on Earth. We all have wings fused to our souls through adversity, fear, and adrenaline, and there is a fellowship that lasts long after the flight suits are hung up in the back of the closet.

Fred Haise and I were both born in 1933 at the peak of the Depression—banks were collapsing and drought turned the American prairies into a dustbowl. Hitler became chancellor of Germany and Franklin Delano Roosevelt was elected the thirty-second American president. During this time, aviation was flourishing in the United States: Wiley Post flew solo around the world; the USS *Ranger*, the first Navy ship built as an aircraft carrier, was commissioned; and North American Aviation and Air France were formed.

Haise grew up in Biloxi, Mississippi, and I came from a military boarding house in Toledo, Ohio, but we both carried the dream of flying. We grew up in small towns and our work ethic and values were shaped by our parents and our community. We were both paperboys, and we enjoyed the Saturday movies and lived for the newsreels. Fred's father was a naval officer who would engage in battle in the South Pacific. Haise played semipro baseball and had an aptitude for sports writing and later, as editor of the *Bulldog Barks* in junior college, he hoped to obtain a journalism scholarship to the University of Missouri. However, his career would not be found in writing. It began as a Marine fighter pilot, winning his wings of gold in the fabled Grumman Hellcat in 1954.

Never panic early experiences come early and often to young aviators. Haise's first calamity arose due to a combination of bad weather and an engine failure, resulting in an emergency landing at a small airport. His was the first McDonnell Banshee jet to land at the Tamiami Airport.

After a tour as a flight instructor at Naval Air Station Kingsville, he decided to become a test pilot. After discharge from the Marines in 1956, he entered the University of Oklahoma to obtain an engineering degree and he resumed flying with the National Guard. While reading the manuscript, I could visualize Haise's destiny: A person who takes risks is free, and the day Haise earned his wings is proof positive that his attitude would carry him very far.

In 1959, he was assigned to Cleveland, Ohio, at NASA Lewis Research Center to engage in zero-G testing. Lewis was the starting point for many aviators who subsequently went into the space program. Three years later, he moved to the mecca of flight research, at Edwards Air Force Base in California, conducting single- and multi-engine tests, and supporting Chuck Yeager's lifting body testing as chase pilot. In 1964, he began a classroom and flying test syllabus at the Aerospace Research Pilot School at Edwards.

Sputnik, in October 1957, began the convergence of our individual destinies. Flight was our life and looking skyward, we saw space as our new arena. The words *higher* and *faster* took on a new meaning for us.

Returning from Korea in 1958, I accepted a position as a civilian flight test engineer on the B-52 at Holloman Air Force Base. Two years later, when I completed the test program, I joined the NASA Space Task Group, supporting Mercury and Gemini. By the time of Haise's selection as an astronaut, I had served as flight director with many of the members of the first four astronaut classes.

In April 1966, Haise was selected for the Fifth Astronaut Class. He spent the first six months in the classroom, visiting contractors, touring NASA facilities, giving speeches, and participating in the field geology- and survival-training programs. Helicopter training and his assignment in December 1966 to the Apollo 2 support crew are some of the events that Haise recounts in detail. Throughout the book, the informal background on other class members often adds humor and provides perspective on the personalities of many of the astronauts I worked with. The story of Bruce McCandless wandering off in the Panamanian jungle to bird watch and his capturing of a deadly fer-de-lance snake, which he hand delivered to the Houston Zoo, offers a rare glimpse of an astronaut the mission controllers worked with on many occasions.

The chapter "Life on the Edge of Space," describes Tom Kelly's challenges and frustration to assure the quality of the lunar lander before it left the factory. I was the flight director for the successful flight tests of the two lunar modules that Haise developed and tested in the Grumman plant. His description of the Apollo 13 oxygen tank explosion as one of his never panic early experiences relays a close and personal sense of the event. My team and I had faced mission crises before—the Gemini 8 emergency reentry to landing in the West Pacific was the closest call we had ever faced. Apollo 13, however, was a matter of survival. It was as tough a test as could be conceived and put to flight control. If there was any weakness, the team would have crumbled. But commitment to one another brought out the fight needed to save the crew.

In 1974, Haise joined the Confederate Air Force, flying a Japanese look-alike aircraft in the *Tora, Tora, Tora* airshow that reenacted the Pearl Harbor attack. I flew this show many times with the Lone Star Flight Museum's B-17 Thunderbird. The show is intense, with eight to ten aircraft wheeling and turning in the smoke-filled sky, over the airfield, accompanied by air-raid sirens, explosions, and narration. Haise, in the Japanese Val bomber, would dive through this melee of aircraft pursued by a single American P-40 Warhawk. An engine failure on a ferry flight triggered another *never panic early* moment when the aircraft tumbled and burned in a crash landing amid a field of cows. Haise received second- and third-degree burns over two-thirds of his body. To contend with his injuries and the pain as he was healing, he kept reminding himself that he was a Marine.

A week after his hospital discharge, he returned to work at the Johnson Space Center and began preparing for the shuttle approach and landing flight-test program. I take my hat off to Haise for including this, because this phase of testing by the ALT crews is often overlooked in space shuttle history.

In 1979, after twenty years with NASA, Haise retired to accept a managerial position at Grumman. He describes his tenure there as purgatory, due in part to the Challenger accident and the NASA/US Congress's disarray in the planning of the space station Freedom program. He considers his work in the business world to be as complicated and stressful as his experience as a pilot and astronaut.

Haise's final chapter, "In the Rocking Chair," addresses our nation's poor international ranking in science and math, and advocates for increased training in science, technology, engineering, and mathematics, to help address the world's future challenges. In order to be part of the solution, Haise joined the board of the Infinity Science Center, a not-for-profit located on Interstate 10 on the Mississippi-Louisiana border. His service is part of his legacy, which ensures that young people are exposed to quality learning experiences on technology.

When I think of Haise's commitment to the Infinity Science Center and its constituents, Václav Havel, a writer of Czech literature and the first president of the Czech Republic, comes to mind. He wrote, "The real test of a man is not when he plays the role he wants for himself, but when he plays the role destiny has for him." I consider Haise a man of destiny.

———

ACRONYM LIST

AC	alternating current	**MCC**	Mission Control Center
ADF	auto direction finding	**MCAS**	Marine Corps Air Station
AFB	Air Force base	**MSC**	Manned Spacecraft Center
ALSEP	Apollo Lunar Surface Experiments Package	**MSFC**	Marshall Space Flight Center
		NAAS	Naval Auxiliary Air Station
ALT	Approach and Landing Test	**NACA**	National Advisory Committee for Aeronautics
ANG	Air National Guard		
APU	auxiliary power units	**NAS**	Naval Air Station
ATU	Advanced Training Unit	**NASA**	National Aeronautics and Space Administration
CDR	commander		
CMS	command module simulator	**NWASI**	Northrop Worldwide Aircraft Services Inc
COAS	crewman optical alignment sight		
CSM	command and service module	**OCP**	operational checkout procedure
CRT	cathode ray tube, more familiarly known as a viewing screen	**OPS**	oxygen purge system
		PGNS	primary guidance and navigation system
DSKY	display keyboard		
EVA	extravehicular activity	**PPK**	personal preference kit
FAA	Federal Aviation Administration	**RASPO**	Resident Apollo Spacecraft Project Office
FSAA	flight simulator for advanced aircraft		
		RCS	reaction control system
GSFC	Goddard Space Flight Center	**SCA**	shuttle carrier aircraft
GTSI	Grumman Technical Services Incorporated	**S/CAT**	spacecraft engineering and test
		SEP	solar electric propulsion
ILS	instrument landing system	**SNAP 8, SNAP 27**	
ISS	International Space Station		systems for nuclear auxiliary power (small nuclear reactors)
IMU	inertial measuring unit		
JSC	Johnson Space Center	**SPS**	service propulsion system
KSC	Kennedy Space Center	**TACAN**	tactical air navigation
LLRF	lunar landing research facility	**TLI**	trans-lunar injection
LM	lunar module	**UHF**	ultra-high frequency
LTA	Lunar Module Test Article	**USAF**	United States Air Force

CHAPTER 1

MY BEGINNINGS

Dr. Burnett had to use forceps to drag me into the world, on Tuesday, November 14, 1933. This resulted in a slightly misshapen head with lots of bruising. My own mother thought I was an ugly baby. She told me that I ate and chewed on everything as a tot—things like rubber bathroom mats, car tires, dirt, and even a pink, infant mouse. Mom wouldn't kiss me for a while following the latter event. My family thought I had a mineral deficiency.

After my birth in Biloxi, Mississippi, we lived for a while in a house at the corner of Howard Avenue and Hopkins Boulevard with Grandad and Grandmother Blacksher. I don't remember her because she died when I was young. Mother told me that I kept asking her to wake up when I saw her laid out for viewing in the dining room. Grandpa Blacksher, whose parents emigrated from Germany, never remarried.

The first house I recall was on Lameuse Street, a few houses north of the famous Barq's root beer plant. Dad had two pointing bird dogs to hunt dove and quail. They were fun to play with, even though Dad thought that playing made them less capable as hunting dogs. My favorite toy was a stuffed brown teddy bear. It was my close companion that I slept with every night. Mom patched up old teddy many times when his seams unraveled from the wear and tear.

Biloxi was a small town of around 14,000 people. It's on a peninsula on the Mississippi Coast, so a lot of activity was centered on the water. We had a wooden skiff with a five horsepower Johnson outboard motor. Dad, Uncle Pat, and I did a lot of fishing in the Gulf, but spent many days in the Pascagoula River and, sometimes, at Horn Island. When the tide was right, Dad and I would hunt for soft shell crabs and flounder at night. We searched sand bars that went out from the seawall for the softshell crabs at low tide, and then for flounder in deeper water as the tide switched. To see the bottom, we used kerosene-soaked rags that were secured to the end of a metal rod—our makeshift torch. The problem with this was that the black smoke smutted one's face and darkened clothes, too. When the Coleman lantern that burned "white" gasoline came out, we had better visibility, so we could work deeper water for flounder and there was no black smoke. Working deeper had a drawback one night, when a small sand shark chased our caught flounders trailing behind us on the leader, but Dad discouraged the predator with a few jabs of his spear. Sometimes wading off the seawall, Dad would cast a net to catch fifty or more mullet for supper. Filleted and deep fried with a cornmeal batter, they were delicious and known as Biloxi Bacon. Biloxi had a large fishing industry with a number of plants that processed shrimp and oysters for distribution around the country. Otherwise, Biloxi enjoyed a small tourist business with a number of hotels.

When I was four years old, we moved to 731 Church Street. Church Street is one block long, ending at Division Street on one end and at Bradford Street on the north end that forms the southern boundary for the Gorenflo Elementary School. At that time, Church Street was a crushed oyster shell road, as were many roads in Biloxi.

Many summers my family went to Otto Mike's Camp up the Pascagoula River for a couple of weeks. It was roughing it. I dreaded the baths in the river or the creek. We stayed in a small cabin with screen coverings on the windows and doors. Air conditioning was unheard of. The Jordans—Uncle Pat, Aunt Irma, and their children, Harry and Rodney—went to Otto Mike's, too. The main pastime was fishing. Dad and I, in our skiff, would go pull up a trotline that we had set across the bayou. Once, when my older cousin, Harry, and I went to run the trotline, he pulled a large snapping turtle into the skiff with us. When it started to act unfriendly, Cousin Harry pulled out his 22-caliber pistol and

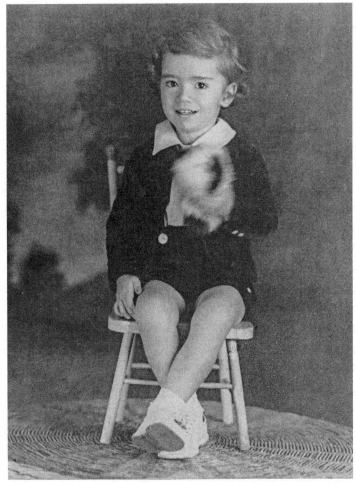

This picture was taken when I was three, in Biloxi, Mississippi. Courtesy of Fred Haise

proceeded to fire away at it. I was impressed by the loud noise as well as the small holes that he made in the bottom of the skiff.

———

On my Dad's side, there were two grandparents, Harry E. and Emma Straub Haise, the latter of whom I called Danny. She was born in 1872 and died in 1956. Dad's older sister, Edith Margaret Haise, lived with her parents (my grandparents). Dad's other sister, Irma C., was Uncle Pat's wife. They lived not far from my grandparents, across an empty lot that led to their house on Delauney Street. Directly across Delauney Street is where my great aunt Neen Abbley lived. She was an old maid whom I often saw while visiting Danny, straightlaced and cold in her bearing. She wore her hair in a bun on the back of her head. She was not big on children, but she seemed to take to me. She fed me generously, including treats like cookies that she baked. Danny made a drink called clabber, which was a preparation of soured milk. It wasn't too bad if laced with a lot of sugar.

Danny was big on molasses with sulfur as a cure-all medication and was quick to feed me some if I showed the slightest symptom of illness. I wasn't sick very much except for a couple of ear infections. The pain was pretty bad, but the cure was even worse. The solution was to pour warm olive oil into the ear and stuff it with cotton. Home remedies were the order of the day. I once stuck a nail in my foot. The prescribed treatment was to beat the wound with a flat "ruler size" board to make it bleed and get the "poison" out. Then the area was generously coated with kerosene. For almost every stomach ailment I immediately got a dose of castor oil. The taste of that stuff could not be disguised by anything. I thought orange juice was the best cover, but it would always float to the top even when thoroughly stirred. I learned early on to stifle any coughs or nose drips.

Grandad and Danny's house was full of the modern conveniences of the time. In the kitchen there was a candlestick phone where you lifted the earpiece off a holder. To make a call, one clicked the holder to get an operator to make the connection. The phone was a party line, so if another household was using the line, one hung up and politely waited. My grandparents had a large, modern-looking refrigerator with a set of coils sitting on top. At my home we still had just an icebox, which was an upright chest with blocks of ice that were replenished to keep food from spoiling.

Grandad lived in a small room at the rear of the house. I spent many hours talking to him about the old days. I had to shout because he could barely hear. He smoked a pipe from the minute he woke up until he went to bed. His fingers were discolored from the Prince Albert tobacco that he bought in two-pound cans. He smoked a corncob pipe that some people called a Missouri

Meerschaum. Grandad said that pipe gave the purest flavor, once broken in. They must have been tough to break in, because he went to great lengths to patch the holes with a mixture of flour and water before throwing one away.

Grandad was born in 1854 in Springfield, Illinois. In elementary school, his class had to memorize a two-verse poem to recite at Abraham Lincoln's final interment ceremonies. After his assassination in April 1865, the funeral train left Washington, DC, and headed to Springfield, making stops along the way in Baltimore, Harrisburg, Philadelphia, New York, Albany, Buffalo, Cleveland, Columbus, Indianapolis, and Chicago. Millions of people thronged the stations to pay their respects.

At Springfield, they loaded Lincoln's casket onto a horse-drawn hearse to carry it to the Illinois State House where he would lay in state, after which he was taken the two miles to Oak Ridge Cemetery. Grandad was a witness to this, and the truly remarkable thing is that, at the age of ninety-five, he remembered the poem. I kick myself for not writing it down. He died when he was just shy of one hundred years old, after falling in the bathtub. He broke his hip and ultimately died of pneumonia.

Edith M. Haise, or Aunt Deedie as we knew her, was a character. She always wore colorful clothing, including impressive hats and large pieces of costume jewelry. She had bright red fingernails and smoked with a stylish cigarette holder. Aunt Deedie was a sales clerk at the Joyce Company clothing store. Sometimes, she would ask me to go get her "cough medicine" at Romeo's small grocery that was half a block down Magnolia Street. Romeo wrote down the purchase in his ledger, because Aunt Deedie never gave me any money. I later found out that the cough medicine was Ballantine scotch.

My mother's family had six girls and one boy, whose name was Fred. Just as with my grandparents, only one of the children in either family finished high school. Mom and her older sisters all worked in the shrimp factories, hiding if the truant officer came by to find out why they were not in school. At family gatherings, when there was a big pile of boiled shrimp as the centerpiece, I always tried to sit by Aunt Thelma. With that shrimp-factory training, she could peel them faster than I could eat them.

In the early 1930s, the Great Depression caused a lot of heartache, but I never felt deprived or poor. Of course, being the only child for seven years and five months assured that I got a lot of attention from my parents. There was no kindergarten back in the '30s, so I started school in the first grade at age five.

Being raised a Catholic, I was enrolled in Sacred Heart Academy. Mom would walk with me for the eight blocks to and from school each day. The catechism was part of the curriculum to prepare me for my First Communion. The teachers were nuns who were strict, but also kind and considerate. During that first year of school, I only got in trouble one time. Another boy and I were caught throwing torn-up pieces of paper down the stairs. For punishment we had to sit in the front corners of the classroom, on a stool facing the wall, for an hour. That stool got really hard and that was a mighty long hour, but I was lucky that my crime had not been passed on to my parents, or I would have been punished a second time when I got home. I consider my early educational experience as one of several major contributors to the discipline that is integral to my character.

My First Communion was a big deal. I was dressed up in a white suit, and all the little girls wore white dresses and lace headdresses. I was nervous when I went up to the altar to receive the host. I was deathly afraid that the host would fall out of my mouth onto the floor when the priest attempted to place it on my tongue and a lightning bolt would come down from heaven to smite me for such a sacrilege, but thankfully all went smoothly.

My sister, Brenda, was born on March 6, 1941. This required some adjustment from me, because I was no longer top dog. As time went on, like any younger sibling, she would get into my stuff. In spite of what she claims, I did not close her up in a closet or push her down the stairs in a cardboard box.

In the second grade, I transferred to public school at Gorenflo Elementary. I loved it, and it was only a half block from home. I ran to school each day and often ran home for lunch. The school had no cafeteria, so we either brought a brown bag lunch or went home. I always had a peanut butter and jelly sandwich with an apple or banana. I've never tired of this lunch, even to this day.

I enjoyed the schoolwork and all of my teachers, except for one. There were two old maid teachers who were sisters. Miss Alma Rich was the principal and very nice. Her sister, Pricilla, was my third grade teacher. She was just the opposite. They were a perfect good-cop, bad-cop team. Pricilla could have played the part of the Wicked Witch of the West in the *Wizard of Oz* without needing special makeup or direction. If one talked without permission, she would poke a hard, bony finger in your chest. Or, if you committed a worse offense, the punishment was to stand in front of the class while she whacked your open hand with the back of a ruler. Thankfully, I only suffered this punishment for a higher crime once. I enjoyed schoolwork and did well enough to be advanced to fourth grade while still in the third grade.

Our introduction to music class was through a rhythm band. The variety of musical instruments consisted of two rulers, a metal triangle with a small metal rod, and a tambourine. After "auditioning" with the rulers, you were assigned to play one of the three instruments. As it turned out, I could not beat the two rulers together and stay in rhythm, so the workaround to keep me in the action was to make me the band leader. I would announce the song at the PTA meeting and then wave a wand that no one paid attention to—the band followed the teacher waving her baton.

In the second grade we had a reading text titled "Freddie the Green Frog." That became my nickname for a while. Then I got the part of a woodpecker in an operetta where all I had to do was dance around a wooden Indian—the teacher knew better than to let me sing. Mom made me a costume with red paper for the head and a stiff, yellow beak that covered my upper body. It was great. I enjoyed chasing shrieking girls around, threatening them with the beak, then dancing around the Indian. I am sure the teacher had second thoughts about casting me for that role, which earned me the moniker "Pecky." That name has stayed with me throughout my life. Most of my friends in Biloxi still call me that, and my nieces and nephews call me Uncle Pecky.

——

With the attack on Pearl Harbor on December 7, 1941, and the start of World War II, a lot of things changed on the Mississippi Coast and in the entire country. The US Army Air Corps established training at the Biloxi Keesler and Gulfport bases, growing to more than 100,000 Army Air Corps soldiers in training. When I walked along Howard Avenue, I had to weave through a sea of khaki. Biloxi quickly transformed from the small, sleepy, coastal seafood town to one that established new businesses to accommodate young military customers. A large number of bars and pool halls sprouted up to provide recreation for the soldiers. At eight years old, I decided to start a business to take advantage of this large group of potential customers who needed their shoes shined. With a shoe box Dad made, rag, brush, and several cans of shoe polish, I was in business. My fee was five cents for a shine. That may not seem like much, but a nickel in those days would purchase five small Tootsie Rolls, an RC Cola, or a candy bar. And four customers would equal a ticket to the Saturday matinee at the Buck Theater, plus a small bag of popcorn. The movie show always included a cowboy picture starring Gene Autry, Roy Rogers, or Hopalong Cassidy; a Merrie Melodies cartoon; and a serial. My favorite had Buster Crabbe playing the astronaut, Buck Rogers. That serial was my introduction to thinking about space. Buck would always be in trouble near the end of each episode. He would often escape death by climbing into his rocket ship and, with the push of a single button, it would lift off with puny, flickering flames emitting from the exhaust. Later in my life, I found out it wasn't quite that simple.

———

Dad joined the Navy when he was thirty-eight years old as a chief petty officer machinist mate. He had twelve years' experience in the Merchant Marine in his younger years, after running away from home at sixteen. He shipped out on freighters, working mostly in the engine room, and traveled to Europe and the Far East. Anyone with boat or ship experience joining the Navy at that time could have gone through the accelerated ninety-day wonder training during World War II to become an officer. The war ended my days at Gorenflo Elementary. We moved to Chicago because Dad had to participate in the commissioning of his ship, YMS-84, a minesweeper.

I had no trouble with my new school or academics, but I had to adjust to the cold weather, snow, and fierce wind coming off of Lake Michigan. Chicago winters were quite a change from Mississippi's warm breezes from the Gulf Coast. During our one-year stay in Chicago, my mother took me to vaudeville shows at theaters downtown. There was a variety of talent, including dancing ladies, singers, magicians, ventriloquists, and comedians, but the only performer I remember was Jimmy Durante. He was the emcee, and I was impressed by his large nose. I also loved when my mother took me to Lincoln Park Zoo.

With the YMS-84 checkout completed, Dad sailed off down the Mississippi River to his next station in Key West, Florida, and we followed. I transitioned into my fourth elementary school and fell right into the rhythm of things, because the basics of reading, writing, and arithmetic were pretty standard across the country. Key West was solid military. The Army had a small detachment and the Coast Guard operated there, but the Navy predominated with a naval operating base, a naval air station, and a submarine pen. I loved riding my bicycle all over the island. My friends and I explored the island with a handmade wooden raft. Once, we came across a large, dead, green sea turtle floating in the water. We towed it to shore with the thought of doing a dissection to save the shell, but the smell made us change our plans and was a sign that I would never become a pathologist. Key West had critters that I had never seen or heard of—one that I learned to stay away from was the Portuguese Man of War. They were easy to avoid because their balloon-like body floated on top of the water and the poisonous tentacles dangled below. I was familiar with jellyfish that stung in the Mississippi water, but they were nowhere near as dangerous as the Man of War. Sand would help cure the pain of a jellyfish sting, but there was no such luck for a Man of War sting.

With my dependent pass, I could go on any base, which meant that I had my choice of four movies every night. Weather permitting, two of them were shown outdoors. All were free except at the naval air station where they charged ten cents. I also enjoyed the USO shows in an aircraft hangar. I saw the Kay Kyser Orchestra and Dorsey Orchestra.

Dad would be out at sea for several days of training exercises on his small, wooden minesweeper. The captain was a lieutenant, junior grade, but sometimes Dad would be in charge, standing watch and remaining on the ship during weekends when it was docked at the Coast Guard base. Several times, I visited Dad aboard. It was really something to play captain and survey the sea of ships with binoculars. I took delight in manning the five-inch gun on the bow. I lined it up on huge ships across the way at the naval operating base by running from one side to the other to turn wheels controlling azimuth and elevation. I suffered a black eye running into a depth charge sticking out from a "Y" launcher used to sink enemy submarines.

When his ship was ordered to the South Pacific for action, Mom, Brenda, and I headed back to Mississippi. To get there, we rode a bus to Miami, followed by several train rides full of military personnel. I was enrolled in Howard II Elementary, my fifth school in five years.

Two things happened when I was twelve years old that had a great influence on my life. One was joining the Boy Scouts as a member of the Dan Beard Patrol in Troop 212. My scout master was E. P. Wilkes, owner of the *Biloxi Gulfport Daily Herald*. The second great thing was becoming a paper boy. It was my small business. Six days a week, I delivered the newspaper.

Each paper boy bought his newspapers for two cents apiece and charged the customer twenty cents for the week. The profit was eight cents per week, per customer. I had Route 16 that started and ended at the *Biloxi Herald* office on Jackson Street. With 170 customers, I made a profit of $14 a week. I was moving up in the world. With all of this wealth, I saved enough to purchase a Cushman motor scooter, which came in handy when I was promoted to office boy. My salary was based on the minimum wage of fifty cents an hour. I also had to learn other paper boys' routes just in case I had to fill in if they were ill, on vacation, or missed a delivery. Freddie Steinwinder was the other office boy, and the two of us covered all Biloxi routes. Freddie had everything east of Main Street and I had all the routes west, to the city limits.

I loved to speed around on my motor scooter, and this led to a few mishaps. One afternoon, I was going too fast on the slick, wet bricks of Howard Avenue, when an 18-wheeler pulled out in front of me. I couldn't stop and had no choice but to lay the scooter down and let it drag me under the truck. It was a close call. The trailer passed over me as I lay on the ground. When I got up, I noticed that a small crowd had assembled outside the old City Hall and the Masonic Temple across the street. Embarrassed by the people staring at me, I quickly got to my feet and sped off on my scooter.

A disadvantage of working after school was that I could not participate in any school sports. I played on a team in a nighttime fast-pitch softball league. With a good arm, I played either short stop or catcher. I might have made the

My graduation photo from Biloxi High School in 1950. Courtesy of Fred Haise

high school baseball team, but football could not have been an option due to my size. I never weighed more than 130 pounds, so I was small. I was also less mature than most in my high school class. Graduating at sixteen, my academic average through high school was a B minus, because I just did not have the passion that a serious life goal would have given me.

I joined Phi Kappa, a high school fraternity that provided a variety of social events, including an annual formal dance. In my junior year I had my first date, but I was afraid to ask any girl that I did not know well, so the logical thing was to ask my next-door neighbor, Rosemary Lizana. I wasn't about to pin a corsage on any girl, so my mom went with me to accomplish that task and then drove us to the dance at the Community House on US Highway 90. Adding to my case of nerves was the fact that I was going to have to dance. My dancing ability never advanced past the two-step my entire life. Despite this, I had fun visiting with my friends.

Through high school, I enjoyed math and science. I also worked on the school newspaper, *Biloxi High Tide,* as the sports editor. That inspired me to get more experience in journalism. The city editor for the *Biloxi Gulfport Daily Herald* gave me an opportunity to cover junior high and peewee league sports events, both during the school year, and hotel conventions over the summer months. Most of my articles were a paragraph of print. When the lead reporter Jack Nelson, future *Los Angeles Times* reporter and later Pulitzer Prize winner, had conflicts, I actually got to cover high school games.

In my senior year, I met Mary Griffin Grant at school, and she became a steady girlfriend. She was one of the few students that had her own automobile. Her dad owned one of the two major drug stores in Biloxi that had a soda bar and booths. Mary and I shared those booths many times before going to the movies just down the street at the Saenger Theater.

In Biloxi, the L&N Railroad track was the demarcation of class. The wealthier families lived south of the tracks and along the beachfront. Mary lived in a large home on the waterfront with a porch, a few homes away from the White House Hotel. She finished school with almost straight As and was the class valedictorian. Each summer, she went off to the Lake Lure Camp in North Carolina. I missed her and wrote letters to her.

———

CHAPTER 2

LEAVING THE NEST
AND LEARNING TO FLY

I moved from home after graduating from Biloxi High School in 1950 to attend Perkinston Junior College, only thirty-five miles away. The school was nestled in a pine forest region, north of Gulfport, Mississippi, on US Highway 49. The school enrollment was approximately six hundred. I earned a half scholarship as the sports editor of the school's newspaper, *Bulldog Barks.* This allowed me to attend all the games from the press box or sidelines as I took notes for my articles. I really didn't feel like I was that far from home because many of my high school classmates attended Perkinston, too. Mary went off to Stephens College in Columbia, Missouri, though we continued to date long distance.

Dormitory life was a big part of my school experience. My roommate in the first year was Jimmy Wilson from Mississippi City. We had a corner room on the second floor of Jackson Hall. Jimmy was an avid outdoorsman who was majoring in biology, which served him well later in life when he had a TV show and news column covering fishing in Florida.

At the nightly open house at the girl's Harrison Hall dormitory, there was always someone at the piano to lead singing, but several scandalous events occurred. The most memorable was when some of the girls hung their underwear out of the windows. Miss Batson was in charge of the dorm and quickly put an end to that fad. The student union was located in the basement of a dormitory building that housed many of the faculty. The primary entertainment there centered on a ping-pong table and a small bar with soft drinks and snacks available for purchase.

For fun, I had a Hallicrafter radio that offered the standard stations and several other bandwidths. It picked up signals from all over the world, including Morse code from ships at sea. Some of the popular radio shows that I listened to were Bob Hope, Red Skelton, *The Shadow*, and *Inner Sanctum*.

Mr. Ware, the band leader and music teacher, was in charge of the dormitory and had a full-time residence downstairs from us. Needless to say, he had a number of issues to deal with over the year. There was the mystery of bottles and cans rolling down the hallway on the second floor in the middle of the night. Jimmy and I were among the primary suspects because we lived at the end of the hallway. We eventually found out that the culprit was Louis Janus who was a straight-A student majoring in physics. For some reason, he would creep down the hallway with a bottle or can and throw it back toward his room. Louis became a naval officer with the submarine service.

Mason "Pie" Thompson was a character from Biloxi who noticed a knot hole in the floor of the vacant room just above Mr. Ware's apartment. He carefully pried it loose and in the middle of the night, dropped a cherry bomb through the hole. Mr. Ware came out of his room fuming.

In my second year, I roomed with Lewis Armistead, who was also from Biloxi. He was a great pool player. Lewis had a partial right hand with one normal finger that he used to his advantage as a good support for a cue stick. Lewis was also interested in the newspaper business and served as my sports editor for the college newspaper.

———

I took advantage of the opportunity to engage in sports in college, since I had missed the opportunity in high school. I went out for baseball and was a backup infielder and an average pitcher. On my best day, my fastball was around eighty miles per hour—that speed is achieved today in Little League. My one-knuckle curve ball wasn't anything to write home about either. I didn't shave for games, so I had a dark shadow of whiskers. When I was on the mound, spectators would make a "Caw-Caw" call, which is why Crow became my nickname among teammates.

One day, I was on the mound against our main rival, Pearl River Junior College, on their field. Perkinston was winning. A tall, lanky, lefthanded batter put one of my pitches over the right field fence. It was a long walk to the dugout for me after that. Fortunately, I was relieved by another teammate who managed to hold on to win, making me feel a little better. At Perk, as we called it, I earned a sports letter that I could not earn in high school. I gave a sweater with my letter to my sister Brenda.

I even played semipro baseball for Mississippi City in the summer. I pitched a game in Mobile, Alabama, against a team that had a fellow named Wilmer "Vinegar Bend" Mizell pitching. He was a lefty and somewhat wild. He pitched a fastball, and it was the first time I had ever seen a baseball appear to shrink the closer it got. I was not a great hitter, but, between the speed of the ball and a lucky impact at the fat of the bat, the ball sailed out between center and right field to the fence. The coach waved me on to try for third base, but I could see that I would not quite make it, so I slid. The spikes in my shoes stabbed into the third baseman's shin bone. He dropped the ball and punched me. It was a win-win—the umpire ruled me safe and the third baseman was thrown out of the game. Vinegar Bend made it to the major league for nine years, pitching for the St. Louis Cardinals, Pittsburgh Pirates, and New York Mets. He later became a US congressman in North Carolina.

In my sophomore year, I became editor of the *Bulldog Barks*. This was my first experience in managing a team to produce a newspaper. Mr. Humphrey, our school librarian, oversaw the operation. I met with the student staff in the library to map out the major stories to be included in the newspaper's four pages. We typed the articles on a Smith Corona or Remington typewriter, and then the text was proofed and sent to the print shop.

The production facility for the *Bulldog Barks* was a print shop in Wiggins, Mississippi, owned and operated by Chester Pratt. He had one noisy linotype machine, which produced lead slugs with a single line of text, the width of a newspaper column. The lead slugs were placed in trays and proofread. This two-step quality assurance was similar to the "double Q points," or checks, by

National Aeronautics and Space Administration (NASA) and its contractors, and were critical steps in testing spacecraft.

Approaching graduation, I hoped to continue my education at the University of Missouri, which was noted for its journalism program. While my B average earned induction into the Phi Theta Kappa Honor Society, I would not earn a scholarship.

This was the Korean War era, and I thought about joining the military to serve my country, just as many of the men in my family had. Dad and my uncles, Walton Dunn and Fred Blacksher, served in World War II, so joining seemed like the thing to do. Dad ended up in the Pacific theater and his ship sunk during the invasion of Borneo, but he survived. Uncle Dunn served in the Marine Corps, also in the Pacific, during the Saipan campaign. Uncle Fred was drafted. Because he was the only son he probably could have gotten a deferment, though he chose to go to war anyway. He served in North Africa and then in Europe. He was killed at Bastogne during the Battle of the Bulge.

My Dad empasized that when I went into the service, I should become an officer. He suggested that I go into the submarine service because he felt that nuclear subs were the Navy's bright future. With two years of college, at the age of eighteen, the best option was the Naval Aviation Cadet Program. Dad did not think much of me becoming a "brown shoe" officer or pilot—a ship naval officer wore black shoes, while naval aviation officers wore brown shoes. Because I was underage, my mother signed my enlistment paperwork.

In October 1952, I checked in at the Naval Air Station, or NAS, in Pensacola, Florida, with some apprehension of what lay ahead. I had never been in an aircraft, and I became even more worried about what I had signed up for when I realized that some of my classmates had flight experience in general aviation aircraft. They used words like ailerons, elevators, and stalls, which were completely meaningless to me.

We were issued Naval Aviation Cadet Program uniforms, and I was assigned to a platoon in Class 52 that was domiciled in brick, multistory barracks. A Marine drill sergeant was assigned to oversee our daily activities. There were two things that were hard for me to master. One was making up my cot so that the cover was taut enough to bounce a quarter. The second was completing my personal toiletries on time to join the daily muster and roll call in front of the barracks. I felt right at home with the community bathroom and shower because it was like being back in the dormitory at Perk.

Our cadet platoon leader Bob Lane from New York made it clear that he was Irish. He even went AWOL one weekend for the St. Patrick's Day Parade in New York and did not get caught—the luck of the Irish.

Overall, I felt that I was in a great situation. I had a nice place to stay, great food, and was paid $90 a month. The gedunk, or snack bar, had great hamburgers and milk shakes.

Ground school gave me confidence that I could actually fly. The curriculum included basic aerodynamics, weather, flight maneuvers, and hands-on experience with simple system trainers for the SNJ aircraft that I would ultimately fly. I learned how to operate a landing gear and wing flaps, among many other system functions. We also received basic parachute training.

We spent a lot of time training in an Olympic-size swimming pool to prepare for the possibility of having to ditch a plane when operating from an aircraft carrier. The hardest physical part of qualification for me, at just over 140 pounds, was pulling a disabled classmate two lengths of the pool with the side stroke. I was concerned about facing the final test of riding the Dilbert Dunker—an SNJ aircraft cockpit placed at the top of a slide, leading into the pool. The cockpit would be released to slide down the ramp rapidly, striking the water with such an impact that the cockpit rotated upside down underwater. The challenge was to unbuckle the harness and swim away from the cockpit before surfacing. A safety diver was there to help those who had a problem extracting themselves. I was pleased that I passed the test on the first run. The Dilbert Dunker was similar to my training in the Apollo program. It was possible, after splashdown, for the Apollo capsule to end up upside down in the Stable 2 position. The capsule could be righted by activating an air compressor to inflate three righting balloons.

Near the end of our sixteen weeks of preflight training, we marched to the supply depot to receive our flight gear. This was a big day. We were given our World War II khaki flight suit and cloth headgear, complete with ear radio receivers and goggles. Bob Lane recognized this auspicious occasion by suggesting we wear our new headgear in our march back to the barracks. This was not well received by our drill sergeant, who made us spend the weekend marching on the parade grounds, rather than enjoying liberty in Pensacola.

My first phase of training was over, and it was time to pack up my foot-locker and move on to the next phase. The program was set up for each trainee to complete basic flight training before proceeding at their own pace through a number of bases in the Pensacola area for advanced training.

I was assigned to NAS Whiting, near Milton, Florida, where I would face the moment of truth—my first encounter with an aircraft. March 27, 1952, was the first day that I would be airborne. It seemed so unreal to me. I reported to a hanger on the flight line and met Lt. Chanaud, my flight instructor. Looking back, I have nothing but respect for those flight instructors who took on the challenge of turning eager, but inexperienced, young men into pilots. He led me through the process of checking my seat parachute and showed me how to adjust the straps for a proper fit. He made a point of identifying the ripcord handle that was pulled to deploy the chute, jolting me into the reality of this flight environment that I was about to enter.

We left the hangar, passing a long row of yellow SNJ aircraft. The *Yellow Peril*, as it was called, with serial number 90829, filled me with awe. Up close, it was much larger than I had anticipated. Some of the aircraft along the flight line had their engines running, making an incredible amount of noise. The SNJ had a Pratt & Whitney R-1340 engine with 550 horsepower. This was a giant step up from my Cushman motor scooter with a 5.5 horsepower engine or my family's 1937 Plymouth stick shift with 70 horsepower.

A pilot must be able to pay attention to minute, intricate details. Lt. Chanaud coached me through a walkaround, preflight inspection of the aircraft—a practice that I would perform for the next twenty-eight years as a pilot. I was taught to look for leaks of fuel or hydraulic fluid, to make sure that the fuel caps were secured, and to wiggle all of the control surfaces, as well as to check the landing gear. Prior to a pilot's inspection, a mechanic or crew chief always performed a more thorough check, recording it on the maintenance "yellow sheet" that's reviewed before going out to the aircraft.

Lt. Chanaud coached me through how to climb on the wing, making sure that I observed the proper strapping in protocol. He talked me through the major control elements that I would be using. This was the real thing, not the static cockpit

My Naval Aviation Cadet portrait, taken in 1952 when I was eighteen years old. Courtesy of Fred Haise

mockup I used in preflight training. This was a machine that smelled of high-octane gasoline, oil, and hydraulic fluids. In later flights, I sometimes smelled vomit that had not been fully cleaned up from a previous student. My adrenaline surged as Lt. Chanaud coached me through starting the engine and called for taxi clearance. The SNJ was a tail dragger, so in order to have adequate forward visibility over the nose, one had to zig-zag back and forth with braking. He let me try this technique when we were safely on a taxiway. Initially, I found it a challenge to prevent over controlling, and my adrenaline flow ramped up even higher because I was actually controlling this flying machine, even if only on the ground.

Once cleared for takeoff, Lt. Chanaud instructed me to add power as he followed through on the controls. This was not only my first takeoff, but the first time ever in an operating aircraft. He told me to let go of everything if he commanded. He nudged the rudder a few times to keep the aircraft straight down the runway. After leaving the ground, I was completely behind in keeping up with the timely raising of the landing gear, reducing throttle, and adjusting propeller pitch. But the world through the plexiglass canopy was mind boggling, like Alice showing up in Wonderland. Immediately below, the houses and autos had grown smaller, but looking toward each horizon, the world seemed to have grown much larger.

The flight lasted one hour and included a tour of the local area. Entering the landing pattern, Lt. Chanaud went over the landing checklist step by step, but with the repeated command to let go of everything if he said so. He engaged a lot of control input to keep us aligned, and I flared properly for my first landing that, to me, seemed to hit hard. This flight changed my life. This simple "box" landing pattern I learned was in the flight plan I prepared for the first landing of the space shuttle *Enterprise* at Edwards Air Force Base (AFB) in 1977. I walked away from that plane knowing that I was no longer going to be a journalist. I wasn't exactly clear where my path would lead, but that was the day I became an aviator.

I received a check ride from a flight instructor named Defreest, who cleared me for solo on my twentieth flight. Then Lt. Chanaud instructed me to fly to Able, which was a large, rectangular grass field six miles away. After I landed, I was to taxi clear of the landing area to park. After climbing out of the aircraft and securing the backseat harness, he told me that I was on my own. On May 12, 1953, I took off feeling the thrill of being totally in control of my destiny.

For the remainder of the time at Whiting, Lt. Chanaud's lessons focused on precision, three-point landings at outlying fields with markings identical to those painted on straight-deck carrier runways. I also learned how to land in crosswinds. With the SNJ, one has to hold a wing down into the wind to negate drift right or left, while applying rudder to hold the nose straight ahead. These landings involved a learning curve in coordinated flight. I could always find a runway in Bagdad, a field that was shaped like an octagon, with suitable cross winds for landing practice. With 69.9 flight hours and eighty-five landings completed, it was time to pack my footlocker and move on to Saufley Field for night and formation training.

Preparing for a night flight involved sitting in the ready room for briefing wearing red goggles. We wore these for several hours to condition our eyes to the darkness. The flashlights we were issued had a red lens, and in the aircraft the instrument lights and the auxiliary lamp were red. Flying at night requires a different set of navigation skills from daytime flying. One major challenge was becoming familiar with an area that looked entirely different in the dark, so you can find your way back home without the use of navigation radio equipment. Because we hadn't had instrument flight training by this time, another consideration was the potential loss of orientation that could occur if one were operating over the open sea at night, facing complete darkness. We kept our flight patterns near areas that were well lit on the ground, even on short cross-country flights. Learning to fly in formation was a lot of fun, and it was exciting to fly so close to several other aircraft.

———

My next move was to Naval Auxiliary Air Station (NAAS) Barin Field—also known as Bloody Barin—in Foley, Alabama. This was where I experienced gunnery. SNJs were equipped with a 30-caliber machine gun to align with a fixed gun sight in the front windscreen. For target practice, a flight instructor led us into a designated range in the Gulf of Mexico. Our flight would start each firing pass from a position above and to the side of the plane that towed a sleeve target on a 1,000-foot line. Following the instructor, we would peel off, one-by-one, from an echelon formation with all aircraft stacked on one side of the leader, to dive and fire on the sleeve target. Each of us followed this sequence until our flight instructor called for the flight to rejoin. The g level in performing the pursuit curve gunnery pass was modest, only reaching 2 Gs. Those flights, or sorties as we called them, were a real hoot, and I wondered if I would ever use these skills in combat. We flew thirteen sorties on gunnery missions. In training, it is a mortal sin to get too close to the tow aircraft, because tow pilots—and I have served as one—do not appreciate getting shot at.

The second phase of training at Barin prepared me to land on aircraft carriers. The runways used for training were painted exactly like the USS *Monterey*, a jeep carrier that I would land on. Carrier landings involved learning to religiously follow the guidance of the landing signal officer who used a set of paddles to guide the final approach for alignment and speed control. A quick crossing of the paddles as one approaches the stern was a command to cut the throttle to idle and make the landing. Disobeying a cut from the landing signal officer could be a court martial offense. On a straight-deck carrier, there could be aircraft parked forward on the deck beyond the capture wires, and one did not want to crash into them. A barrier fence was normally up to trap the aircraft, should its tail hook not catch any of the capture wires. Today's modern carriers have a canted deck so the pilot can add power to go around for another approach.

After completing fifty-eight field carrier-landing practice landings, I was cleared for the ship. On September 10, 1953, I launched from Barin Field, in a flight of four, led by an instructor. We went by the starboard side of the carrier and broke one-by-one overhead for my first official carrier landing. My adrenaline flow was at a new level. A docked aircraft carrier looks huge, but from the air looks incredibly small. After six carrier landings on the USS *Monterey*, I became a newly qualified carrier pilot.

My first experience with the Link Trainer, a live, operational simulator, took place at NAAS Corry Field, in Pensacola, Florida, where I would also receive additional instrument training. The Link Trainer did not duplicate the true handling qualities of a real aircraft, but was excellent for improving one's grasp of instrument approaches.

For actual flights in the SNJ, I rode in the back seat. After airborne, I pulled a cloth hood over my head, which was called going "under the bag." The purpose was to block the view outside the aircraft. I experienced vertigo for the first time: After a turning maneuver, although the gyro horizon instrument indicated a level orientation, I felt like I was still turning. This sensation is caused by the positioning of hairs in the middle ear during the turn. Understanding this effect is critical because a pilot has to develop the discipline and conviction of believing the flight instruments. I logged 13.8 hours in under-the-bag instrument flight to complete the basic flight training syllabus on October 5, 1953. Over the one year since reporting for duty, I had flown 198.7 flight hours. I felt confident and at home in the air as I packed up and left Pensacola. I hitched a ride with a fellow cadet to report to NAS Corpus Christi, Texas, for advanced flight training and my next assignment.

I arrived at NAS Corpus Christi and was directed to see a yeoman for my assignment. He asked me if I was a Navy or a Marine Corps cadet, and I answered Navy. Assignments were made randomly and were determined by when one showed up. He opened a logbook on his desk and told me to report to an Advanced Training Unit (ATU) for seaplanes. He certainly wouldn't have made that choice if he noticed my demeanor and steely eyed look of a budding fighter pilot. Right then and there, I decided to declare that I was a Marine cadet, because they were assigned to fighter or attack training. I knew my father would be even more displeased to see me move from his Navy and the sea to become a US Marine. The yeoman directed me to another office to see a Marine major. He told me to report to NAAS Kingsville in Texas for all-weather flight and fighter training in the Grumman F6F Hellcat, a famous fighter plane during World War II—it was noted for shooting down more Japanese aircraft than any other US fighter plane. I was walking on air.

At NAAS Kingsville, all-weather flight, anti-submarine, and fighter training were conducted at South Field. It had black asphalt 5,000-foot runways, which were suitable for the Beechcraft SNB, General Motors TBM, and Grumman F6F aircraft. North Field, with 8,000-foot runways, supported jet training in the Lockheed TV-1 and TV-2 aircraft.

In ATU 801, I received a few weeks of ground school to prepare me for instrument flight training in the SNB. This included an introduction to instrument approaches, like basic aural range, auto direction finding (ADF), instrument landing system (ILS), and Ground Controlled Approach. We also covered flight planning, filing a flight plan, and interpreting an approach plate—a chart that covers instrument approach.

The SNB was powered by two Pratt & Whitney 450 horsepower R-985 engines with a tail wheel manufactured by the Beech Aircraft Company. It was also known as the twin-engine bug smasher. The aircraft was flown from two seats up front. The student pilot flew from the left seat and the instructor, the right. There were seats in the rear available for five passengers.

Two students were assigned for each three-hour flight and received an hour and a half of instrument flight training. To assure dependence on instruments, several metal pieces were installed to shield the outside world. The maneuvers and techniques we learned were difficult. My first instrument approach was performed using an aural range, which required listening to the Morse Code signals for A or N to ascertain whether you were off the aural beam center, and which way to turn. A steady tone meant you were on the beam. Flying on instruments is a challenging undertaking that, if not carried out properly, could result in accidents. At my experience level, handling the multitasking requirements of instrument flight was the most difficult that I had encountered. On November 18, 1953, I received a stamp in my logbook that I had completed twenty hours of training on twenty-three flight sorties, with an average mark of 3.25 out of 4.00, and was qualified for a standard-type instrument rating.

Another big step up on my journey was training in the Hellcat. I attended ground school and spent some time sitting in the cockpit to become familiar with all of the instruments and switches that seemed a quantum leap from the previous aircraft in basic training. The F6F Hellcat was triple the weight, had more than three times the horsepower, and a maximum speed twice the SNB's. The Hellcat also had 50-caliber machine guns, a tail hook, folding wings, engine cowl flaps, and a two-stage supercharger for the Pratt & Whitney R2800 twin Wasp engine. The latter capability allowed flights above 20,000 feet, which required using an oxygen mask. And the Hellcat only had a single seat, so being alone for the first flight was new and exhilarating.

I was assigned to a flight of six students with Monty Harouf, our flight instructor. We were in awe of Monty as he had flown the F6F off carriers during World War II. Another member of our flight was a French cadet. When weather delayed flying, we often passed the time playing Hearts. In that game, if at all possible, one had to avoid receiving the queen of spades, known as the Black Witch, because that was a thirteen-point penalty. Our French cadet enjoyed laying the Black Witch, even if it messed up his own hand. I think it was a byproduct of the fatalism he developed from fighting in the First Indochina War.

Our flight instructor imposed financial penalties for any error or oversight. They varied from ten dollars for such things as not opening up the engine cowl flaps after landing, to one hundred dollars for forgetting to lower the landing gear. We were encouraged to note errors made by other members of the flight. On one flight, I noticed a friend's aircraft had two red cylinders that we called beer cans sticking out of his wings. The cans were an added safety measure to prevent the wings from folding. Both my friend and the plane captain had missed this before leaving the flight line. I hesitated before calling it out on the radio, because I knew this would bring a big fine, but better a few dollars than possibly losing a life. The money collected was to be used for a party to celebrate the end of this phase of training.

We built proficiency in formation flight and simulated air-combat maneuvering, which entailed two aircraft flying toward one another head-on, then turning to engage in a series of high-g turns. The winner was the one who maneuvered the best near the edge of stall to shorten the circle and get in shooting position behind the adversary. Weapons training included bombing attacks with small, water-filled "bombs" that emitted smoke on impact. Air-to-air gunnery involved an aircraft towing a canvas sleeve target.

Toward the end of this training segment, we flew a cross-country flight to El Paso, Texas, in a loose formation at a fairly low level, changing who was leading several times based on the flight instructor's call. We navigated without any electronic aids, by dead reckoning, or looking at and paying attention to landmarks on the ground. Water towers with city names, railroads, and airports were good references.

I completed F6F training after logging 80.7 hours and received orders back to NAAS Barin Field in Alabama. I took my final test to earn my Navy wings of gold after training, for the second time, to land on the USS *Monterey*, but this time with the F6F. It was a bright, sunny day, and the sea was calm. The first pass went well. Next, I heard the booming voice of the air boss informing me to get my canopy opened, which would help to hasten escaping the aircraft if it landed in the water. The critical thing in the carrier approach was for the aircraft to be maintained just above the stall speed to make sure it landed after the landing signal officer signaled the cut, so the plane would not float past all the arresting cables. In rapid succession, I completed five more landings and

Proudly receiving my Navy Wings of Gold in 1954 with my sister Brenda.
Courtesy of Fred Haise

received one catapult shot off the deck. This completed my flight training. With 307.2 total flight hours, I earned my commission as a second lieutenant in the US Marine Corps and received my Navy aviator wings. Another classmate and I were the first in our group to complete training.

—

CHAPTER 3

INTO THE JET AGE

With my commissioning, I received a uniform allowance so I purchased a wardrobe of Marine uniforms. I was also financially solvent enough to buy my first automobile—a 1956 Chevrolet two-door sedan. I headed back to NAAS Kingsville to start jet training at North Field. I lived in the Bachelor Officer Quarters, which were a step up from my cadet days. I now had a private room with a shared bathroom.

I began preparing for my training in the Lockheed TV-1 and TV-2 jet aircraft. I was issued new flight equipment that included a helmet, complete with goggles and an oxygen mask, since most flights would be above 10,000 feet. To acclimate us to oxygen deprivation, we went through training in a hypobaric chamber, so we could be familiar with what hypoxia feels like. The chamber operators and a flight surgeon observed through windows. We were taken up to 20,000 feet and told to take off our oxygen masks. The symptoms of oxygen deprivation vary between individuals in their onset and intensity. My only symptoms were that my fingernails turned blue and I felt a tingling sensation throughout my body. What is so insidious is that one doesn't immediately experience any breathing difficulties. I was also trained in the use of the Martin Baker ejection seat that offered a plan B escape from the aircraft in an emergency.

After an aircraft handbook exam and some cockpit time, I was ready for my first TV-2 flight. As I strode along the line of sleek, shiny aircraft, I was reminded of my childhood hero, Buck Rogers. I noted the shrill whine of the jet engines and the smell of kerosene. The TV aircraft was equipped with an Allison J33 centrifugal compressor turbojet engine that provided 4,600 pounds of thrust. At the time, it seemed like so much power, but it was a long way from the Saturn V rocket that I would fly sixteen years later, where each of the five engines on the first stage produced 1,500,000 pounds of thrust.

On my first flight, I found the TV easy to maneuver. I was impressed by the large, clear canopy that, on a clear day, from 25,000 feet altitude, allowed me to see that the horizon went out to over fifty miles. The curriculum included formation, air-combat maneuvering, instrument flights, cross-country navigation, but no weapons training.

I now had the experience of flying four different types of airplanes. I realized that aircraft, like people, had different characteristics, known as handling qualities. After 21.7 jet flight hours, I received the following statement in my flight logbook: "The above aviator has completed the prescribed Jet transitional training with no accidents or flight violations during this period."

Now, I was off to the big time with orders to report for duty with VMF-533, an operational fighter squadron in Marine Air Group 24, at Marine Corps Air Station (MCAS) Cherry Point in North Carolina. I would be flying the McDonnell Douglas F2H-4 Banshee. It was the last of the series produced. It had two Westinghouse J34 engines providing 3,600 pounds of thrust each. It was also equipped with four 20-millimeter cannons and eight weapons pylons under the wings to carry bombs or rockets. I found the aircraft easy to fly, both at low speed and high speed. Within a few flights I felt comfortable with the Banshee, which produces a shrill shriek from the two engines.

———

I lived in the plush Bachelor Officer Quarters, which included a cafeteria. I often attended the Friday night gatherings at the Officers Club with fellow pilots in the squadron. The first person who I got to know there was a fellow second lieutenant, whom we called "Gravy," in lieu of his last name, Graves.

The Korean War ended with the Armistice, signed on July 27, 1953, which meant that there was no longer a need to have as many pilots. Some of my class took advantage of leaving the flight program after only serving two years. Many squadrons had 30 percent more pilots assigned than needed. For the pilots who remained, there were not enough proficiency flying hours allocated, which resulted in several accidents. My friend Gravy lost his life due to bad weather while returning from a cross-country flight. It was my first experience over the years wearing a black band at a memorial service for those lost in aircraft accidents.

Even though my girlfriend from Biloxi, Mary, had gone off to college in Missouri, we had maintained our relationship. After I completed Naval flight training and was commissioned into the Marine Corps, Mary and I got married in 1954. I think becoming a commissioned officer convinced Mary's dad that I was acceptable as a son-in-law. So, Mary, with her parents and sister Susan, came to meet me in New Bern, North Carolina, where we wed at a Methodist church. Mary and I left the ceremony for a brief honeymoon in Washington, DC. We initially settled in base housing at MCAS Cherry Point in North Carolina and added Mr. Chips, a fox terrier, to our household. Shortly thereafter we moved to a duplex in Morehead City that afforded walking distance to the water that we both enjoyed.

My best friend in the squadron was Kenneth L. Ford, who was known as "Tex." He had one number lower than my serial number 064534 and liked to remind me, from time to time, that he was my senior. We enjoyed hunting snakes in the thick honeysuckle bushes in the North Carolina woods. Often, we would crawl through the dense foliage. Once, Tex came upon a copperhead and jumped straight up through the vines to avoid getting bitten.

Our squadron went for maneuvers for six weeks in Puerto Rico, where we lived in tents adjacent to the flight line. There was an outdoor open-air shower that accommodated ten, with canvas walls for privacy. Tex and I carried our rifles with us for the trip, so we could bag some large pythons or boa constrictors, which did not materialize. This dream turned out to be a good lesson to always do your homework. A check of Puerto Rico in the *Encyclopedia Britannica,* as Google wasn't around yet, would have informed us that the mongoose had been brought in to clear the island of snakes. Since Tex and

I were adaptive, we overcame our disappointment by hunting the fast, tricky mongoose around the base perimeter.

Our operating base in Puerto Rico was Roosevelt Roads Naval Station, in Ceiba, which was built in 1943 during World War II. The primary purpose was to furnish training with live ordnance on the nearby Vieques Island range. I experienced the fun and excitement of dropping bombs, shooting rockets, strafing, and air-to-air gunnery. The six-week exercise allowed for lots of good flying, and it was the first time that I dropped 500-pound bombs or fired five-inch, high-velocity aircraft rockets against targets. I also got to exercise the Banshee's 20-millimeter cannons against a towed sleeve target.

My first real *never panic early* situation happened during a routine cross-country flight. This mantra was something I picked up from my time at Edwards AFB, as taking action abruptly in an event could cause the loss of a valuable, possibly one-of-a-kind aircraft. I was with Tex, and we were headed to MCAS Marine Masters in Miami on September 1, 1954. Scattered thunderstorms were predicted, but, as we approached Miami, we found ourselves in the top of a storm at our 41,000 feet cruising altitude. Tex was leading and signaled for speed brake deployment and power reduction for the letdown. I delayed in following his throttle reduction. My rapid throttle movement to avoid overrunning resulted in the flameout of both engines and a surge of adrenaline. I could not stay with him and Tex slowly disappeared from view. Not knowing about other air traffic below, I made a turn toward the Atlantic Ocean, using a small wet-compass that was hard to read with the turbulence. I felt fine when I stopped the turn and the attitude indicator showed that I had my wings perfectly level, but that feeling went away because the airspeed was rapidly increasing and the altimeter was unwinding—I was in a graveyard spiral. While the attitude indicator showed that the wings were level, it was actually frozen because there was no alternating current power coming from the dead engines.

Quickly, I referenced the needle-ball instrument to level my wings and calmly began to maneuver. I made a roll-control input to center the needle vertically and engaged the rudder pedals to force the ball in the attitude indicator to the center. It was similar to a carpenter's level. I knew the wings were level, but I didn't know if I was right side up or inverted. With the altimeter still unwinding, I moved the control stick aft to pull some G, to effect the pullout from my spiraling dive. Passing 20,000 feet, I started the engines and continued easterly out over the Atlantic Ocean, clear of air traffic. After descending low enough to be beneath the clouds, at about 1,500 feet, I turned back westerly. I called Marine Masters tower for an ultra-high frequency (UHF) DF steer to the field, because my single ADF for navigation was useless during thunderstorms. Naturally, the way the day was going so far, the tower called to inform me that their equipment was out of commission. My next concern was fuel, which was down to nearly 800 pounds—500 pounds is considered an emergency.

I spotted an airport, which was like finding an oasis in the desert. It turned out to be private, with 5,000 feet of runway that looked much longer to me. I made my landing approach, just clearing a building near the end of the runway, and chopped the power to drop in hard, blowing the right tire at touchdown. The plane veered off the runway and got caught in the sod and mud, which resulted in a quick stop. I found out that I had landed at Tamiami Airport, near the Tamiami Trail. It didn't surprise me that I was the first ever to visit in an F2H-4 Banshee.

My initial recovery plan was to load up with just enough fuel to get to Marine Masters. With this light load, I would go to full throttle, holding the brakes at the very end of the runway. After releasing the brakes, I planned to lower the flaps, approaching 90 knots, to assist getting airborne before the takeoff end of the runway. But before executing this plan, a Marine helicopter dropped in with instructions to take me to visit with the commanding general at Marine Opa Locka. Needless to say, I was not looking forward to this meeting, but I popped into his office, came to attention with a crisp salute, and was invited to have a seat. The general was in charge of the Opa Locka operations that supported only propeller aircraft. He told me that he had never flown jets, but offered a plan B of towing the aircraft to Miami International, which was twelve miles away.

In thinking back, I guess that this wise general decided to test this young, wet-behind-the-ears pilot's judgment, and if I had not opted to have the aircraft towed, he would have decided for me. I borrowed a wheel assembly from Tex's plane, so mine could be slowly towed by a tug vehicle. I suffered an embarrassing ride through the streets of Miami. There were lots of people along the way waving, but I did not feel like a triumphant Roman general.

It was a blessing that the amount of flap damage did not warrant the incident being declared an accident. Despite the dire possibilities, this flight ended well, all things considered. The combination of pilot error, diagnosing the problem, adapting to the situation to prevent a catastrophe, as well as the blind luck of stumbling onto an airfield, taught me valuable lessons. But it seems as though trouble has followed me, leading to many adventures through my life, and, somehow, I have been blessed to recover.

Of course, this episode did not sit well back at MCAS Cherry Point, because an article in the *Miami Herald* misspelled my name—it was published as Hays. My squadron commander's name was Col. Hays. Needless to say, this caused a lot of confusion.

In January 1955, I transferred to VMF-114, a sister squadron that flew the F2H Banshee and was upgrading to the Grumman swept-wing F9F-8 Cougar, which had supersonic capability in a dive. The Cougar had a Pratt & Whitney J48 turbojet engine that produced 7,250 pounds of thrust. Typical of the aircraft manufactured by the Grumman "Iron Works," it felt solid just sitting in the cockpit. It also was a stable and sturdy aircraft in flight, with a g limit over seven. On April 9, 1955, I climbed aboard an F9F with the serial number 134236 in order to experience breaking the sound barrier.

I climbed to just over 40,000 feet, rolled inverted, and pulled through to enter a steep dive. Only later would aircraft have sufficient engine thrust to go supersonic while in level flight. I found my short supersonic experience unimpressive—there was no buffeting, no wing roll or rocking, no sound change—just the airspeed meter increasing until it read Mach 1.2. But I surely felt good to be moving up to supersonic jet jockey.

———

After 79.1 hours of weapons training, I felt combat ready, except the Korean War was over. Throughout my life I've felt that I missed out, and even felt a little guilty, for never having been in the harm's way of battle. Many of my friends whom I served with in the reserves, in NASA, and in NASA's Astronaut Corps had seen combat in Korea or Vietnam.

My next orders were to report, once again, to NAAS Kingsville. This time, I would serve as a flight instructor. Tex also received orders to the same ATU 801. With our wives, we settled into a duplex home—two duplicate homes joined by a joint wall—in the community. Within the year, Mary and I had our first child, Mary Margaret Haise.

Just as with the F6F training earlier, I had become Monty, the trainer versus the trainee, and was assigned to oversee a flight of several students. The curriculum consisted of formation flying, instruments, simulated weapons delivery, night flight, and navigation. I checked out in the T-28B Trojan that had replaced the F6F, and it was a relatively high-performance propeller aircraft, though lacking the great wartime record of the F6F. The Air Force utilized the T-28A that had less performance. It had a two-blade propeller and an 800 horsepower Wright engine. The Navy version had a three-blade propeller and a more powerful 1,425 horsepower Wright R-1820 engine. I had some sorties where I led the flight of students to altitudes approaching 20,000 feet, in the unpressurized cockpit, to give them experience using their oxygen masks. The aircraft was rated at a service ceiling of 35,000 feet.

This era of my professional life was one of the most enjoyable because I flew a variety of missions and clocked many hours of solo flying time; I averaged eighty flying hours a month. I experienced fear one evening, however, when I led four students in night flying. As we approached the airfield, I noted that a layer of low clouds had moved in from the Gulf of Mexico, at 3,000 feet. I called for my flight to hang tight, while I let down slowly to punch through the thin cloud layer. As I broke through, I looked left and right, and all but one student had disappeared. I spent the next thirty minutes tracking them down one-by-one, like a nervous mother hen rounding up her chicks. Fortunately, they were all accounted for.

Our sister ATU operated the Grumman S2F, which was the first aircraft developed specifically for anti-submarine missions. To qualify their pilots as all-weather instructors, I moved back to flying the SNB, carrying a double student load of cadets and pilots from the submarine unit. The curriculum was similar to what I was taught as a student. Lessons in basic instruments were followed up with a variety of instrument approaches. We also took a small step into the future when we utilized ADF. In the civilian world as well as in the military, planes were equipped with ILS capability.

I took pride in training my students to learn the challenges of instrument flight, making sure that they would pass their check ride with another instructor on their last flight of the curriculum. I wanted them to feel pressure so they would continue to work at improving their abilities, so there were several tricks I used to keep them from becoming overconfident. Instrument flight in an approach requires multitasking, and the pilot is required to make adjustments in the aircraft such as deploying landing gear and flaps, while answering to the Federal Aviation Administration (FAA) radio station, ground control, or the tower. So, to instill in my students that flying the aircraft was the highest priority, I told them that FAA controllers got a notch in their lead pencil for every pilot they led astray (that's not true). I instructed them to just say "stand by." Some instructors screamed at their students in frustration when they were not performing well, but I silently just held my breath.

I came up with many ways of creating distractions in the cockpit that kept my students alert. At the right time, for effect, I would drop a pencil and while retrieving it turn off the fuel valve on the deck for one engine. The engine would continue to run for a bit from the fuel in the lines and then suddenly sputter and quit. The student had to handle the shock of losing an engine and respond accordingly. During night-flying missions, I slid my window open just a bit and secretly held an index card in the gap. The flapping and shredding of the paper sounded like a structural failure. As a result, I never had a student flunk their final check ride.

———

CHAPTER 4

BACK TO SCHOOL
AND INTO NASA

Before my discharge in August 1956, when I was twenty-two years old, I wrestled with what I should do next. I was married with one child and my resumé read two years of college and 1,734.2 hours of flying time, so I decided to become a test pilot. From my limited research, I knew that I would need a degree in engineering. I wanted to attend a school where I could continue my flight experience flying high-performance jets in a military reserve unit. That narrowed my options down to the University of Oklahoma, where the Air National Guard (ANG) had an opening at Oklahoma City.

The ANG income, coupled with my G.I. Bill funds and affordable school housing, had us financially stable. Since I had lived in Texas, the university administration wanted to charge an out-of-state fee. I argued that since I was a member of the Oklahoma ANG and could be called into action by the Oklahoma governor at any time, the out-of-state fee should be waived. I won the argument, which further improved our financial position.

We settled in Norman, Oklahoma. Mary and I were among many other student residents around the same age, living in the eight converted World War II barracks. Most of us had young children. Since I had served in the military for four years, I felt older than most students in my classes. I was definitely past the frat days, but Mary and I had a group of friends. We often had outdoor gatherings to make homemade ice cream with the old, manual-crank container. Our second child, Fred Thomas Haise, was born during this time. Since I was a junior, we decided to give him the middle name, Thomas, rather than have him be a third.

———

One day, there was a loud knocking on our door. Our neighbor was brimming with excitement as he informed us that the Russians had something orbiting the Earth making beeping sounds. This was the flight of Sputnik on October 4, 1957. This small satellite opened the space age.

Several complications arose in my processing to join the Oklahoma ANG 185th Fighter Interceptor Squadron. First, I resigned as a Marine and my officer's commission had to be changed to a United States Air Force (USAF) commission. Second, my Navy Wings had to go through a paper review to declare them comparable to Air Force Wings before I could start flying the P-80 *Shooting Star*, the first US operational jet fighter. I felt right at home in transitioning to the aircraft because it was similar to the Lockheed TV-1 that I had flown in training. Between my flying sorties on the monthly weekend drills and an annual two-week training camp, I was surprised to find I averaged more flying hours each month than I had on active duty. The P-80 was a mature aircraft that had a relatively simple cockpit layout and systems. For example, the caution and warning indicators consisted of two small lights, one red and one orange, protruding from the instrument panel. On one night flight, I was startled when the bright red light came on, signifying the aircraft was on fire. Following the adage to *never panic early*, I quickly scanned the instrument panel for the exhaust gas temperature reading to find that it was normal. I then called my wingman to check for fire or smoke trailing the aircraft. He reported that things looked normal, so I unscrewed the cover on the warning light and removed the bulb to continue the mission.

In July 1958, our squadron transitioned to the supersonic F-86 Sabre jet. It was the first aircraft I had flown with an afterburner that could increase the thrust of the GE J47 engine from 5,550 pounds to 7,650 pounds thrust, which caused a

The 185th Fighter Interceptor Squadron of the Oklahoma ANG. I'm third from the left in the middle row, with my hands on my knees. Courtesy of the Oklahoma Air Guard

pronounced kick-in-the-pants feeling. I was given an assignment to pick up one of our new aircraft at the USAF Sacramento Depot in California. Since I planned to just fly out and come straight back to Oklahoma in the new plane, I wore my flight suit, with a back parachute, and carried my helmet bag. I climbed the stairs to board the TWA airliner at Will Rogers Airport, and the flight attendant took my ticket and asked if she could store my parachute in the coat closet. I thanked her but told her I would prefer to keep my gear. Remembering that experience makes me realize how much airline travel has changed.

I had one *never panic early* experience in the F-86. I experienced a primary hydraulic system failure in flight, and recycling did not restore pressure. After I landed, when the crew chief climbed the ladder up to the cockpit, I told him of the problem. I then attempted to recycle to demonstrate to him that it didn't work. Shortly thereafter, the fire warning light came on and, in my cockpit rearview mirrors, I could see smoke pouring out of the aft end of the plane. The crew chief moved away and, after stop-cocking the throttle and shutting off the engine master switch, I quickly followed. The firemen arrived with foam spray, but were unsuccessful in saving the airplane.

———

It was easier and more interesting to study aerodynamics and aircraft structures because I had firsthand flying experience. Several of my classmates were on active duty with the Air Force. One of them was Mike Adams. He was later selected as one of the astronauts for the Manned Orbiting Laboratory, and when it was canceled, he moved over to fly the X-15, while I was a research pilot at NASA Flight Research Center.

I enjoyed meeting with my classmates to work problems, often using our slide rules. I still have my Post Versalog. It needed an occasional dose of talcum powder to keep it from sticking. I could always tell an engineering student because they had a slide rule in a leather case hanging from their belt. The university had a mainframe computer that gave us a rudimentary background in FORTRAN, or programming. Input/output to the computer was made with punch cards held in trays.

An advantage of being a student was having a pass to attend athletic events. I saw some remarkable football under coach Bud Wilkinson during my three years at Oklahoma University. I witnessed when OU lost the game to Notre Dame that ended the still-existing record of a forty-seven game winning streak. I met wrestler Danny Hodge and subsequently became interested in college wrestling. I had never seen or heard of it growing up in Biloxi. Danny was undefeated in his career at OU with forty-six wins.

In my last year of college, I contemplated my next career step. Stanley Newman, my squadron commander, steered me toward the National Advisory Committee for Aeronautics, also known as the NACA. He told me that I would have the opportunity to test many types of aircraft in the NACA, which shortly after became NASA. Taking Stan's advice, I flew cross-country flights to NASA Langley Research Center in Virginia, then to NASA Ames Research Center in California. Neither had any research pilot positions open. I was similarly informed at Ames that no openings were available at the Flight Research Center at Edwards AFB in California, where the X-15 program was just getting underway. At NASA Lewis Research Center in Cleveland, now Glenn Research

Center—named after John Glenn, the first US astronaut in orbit who also became an Ohio United States senator—I was advised by their chief pilot, William "Eb" Gough, to file an application. I was hired on September 14, 1959, as a GS-7 aeronautical research engineer and pilot at a $5,430 annual salary.

For all of our prior moves, Mary and I had rented a U-Haul truck and packed up everything ourselves. But now, with government orders, we were blessed with the assistance of a professional mover who had an 18-wheeler and workers to pack everything in boxes.

—

Flying into Cleveland Hopkins Airport, a hangar stood out on the north side of the airfield, with the large letters NASA painted on the curved roof. I could tell that it was a new paint job that covered up the old letters, NACA. The Lewis Research Center spread out behind the hangar in a number of facilities. Several wind tunnels tested jet aircraft propulsion systems. A rocket engine test facility was located in the Rocky River Creek. The Engineering Research Building had a number of small cells that were used to test lubricants, ball bearings, and other items. There was also an auxiliary facility for remote testing at the Plum Brook Station that included a large vacuum chamber. It has subsequently been used for the vacuum chamber testing of the *Orion* capsule.

After looking over the area, Mary and I decided to rent a home in Berea, a small town not far from Cleveland Hopkins Airport. Baldwin Wallace University was located there, and it offered an easy commute to a rear entry gate to Lewis Research Center.

In conjunction with this move, I arranged a transfer to the ANG unit at Mansfield, Ohio. The 164th Tactical Fighter Squadron was equipped with the Republic Aircraft Company F-84F fighter-bomber. The swept wing Thunderstreak was equipped with a Wright J-85 engine that produced 7,220 pounds of thrust.

Eb Gough, reserve Navy captain and my new NASA boss, showed me to my desk in the Pilot's Office. Eb was an experienced aviator and looked the part of a Navy captain. He had a gruff voice and also a tremendous memory for detail. Jokingly, we said that if you asked Eb for the time of day, he would tell you how the watch worked.

I couldn't have asked for more in my workstation. I had a great view from the second-floor windows of the hangar, looking out across the airport toward the Commercial Aircraft Terminal and a large building on the southwest side that was known as the Cadillac Plant. This huge building had supported large-scale production of World War II vehicles. I felt honored to join a pilot group of five. There had been five, but Bud Ream had just left for NASA Langley Research Center, and I was essentially taking over his "still warm" seat. Other pilots were Eb Gough, Bill Swann, Joe Algranti, and John "Jack" Enders. Jack was a former Air Force pilot. Eb and Joe were former Navy aviators, and Bill Swann was never in the military, but trained at Parks College in St. Louis. NASA Lewis had been the starting point for several other pilots who I joined up with later in my career. Joe Vensel, Joe Walker, Neil Armstrong, and Warren North all had transferred to other NASA centers from Lewis.

When I later transferred to the NASA Flight Research Center, Joe Vensel was head of Flight Operations and Joe Walker was my boss in the Pilot's Office. When I joined the Astronaut Office at the Manned Spacecraft Center (MSC), which is now known as the Johnson Space Center (JSC), Warren North headed the Crew Training and Procedures Division, and Joe Algranti headed the Aircraft Operations Division. I served on the backup crew with Neil Armstrong for the Apollo 8 Mission and as the backup lunar module (LM) pilot for his Apollo 11 Mission.

I was impressed with the array of aircraft on the hangar floor. I saw a Navy R4D and a USAF C-47 that served as our transport aircraft for personnel and cargo. The Air Force had offices on the first floor of the hangar, facing the runways. A colonel was in charge of several personnel to oversee and observe specific aircraft engine tests of interest to the USAF. There was also an F2H-2 Banshee, a B-57A Canberra, and a B-57B Canberra. A small North American L-17B *Navion* aircraft was nestled in among the larger ones. This Army liaison aircraft was used for short flights with personnel and had short field capability.

Quickly, I started preparing for checkouts of the various aircraft. I followed the normal script of studying the handbooks, often while sitting in the cockpit of each particular aircraft. When I felt I had a handle on the normal and emergency aircraft systems and procedures, I would schedule a flight checkout. One of the other pilots would lead me through the normal preflight checks, to the point of starting the engines. Except for the transport aircraft and L-17B, I was on my own. To back me up if something came up during the flight, the pilot who had conducted the checkout would be on standby at the radio in our office.

My first flight in the R4D was a local flight that included various maneuvers, instrument approaches, and a number of touch-and-go landings. I found the plane to be easy to trim but heavy on the controls, particularly the rudder control—I had been spoiled by flying more nimble aircraft. I made several instrument approaches, using the various navigation aids such as the ILS. This was the first time that I had seen the ILS capability and I was impressed that it had me aligned with the runway when I looked out several hundred feet above the ground.

———

Project Mercury was the nation's first manned space program. Bob Gilruth led the Space Task Group at NASA Langley, and many personnel transferred from other centers to fulfill the staffing requirements. A number transferred from Lewis Research Center to Langley to join this newly formed group and a few went to NASA headquarters. Until they could sell their homes, an aircraft ferry operation was conducted on Mondays and Fridays in the R4D. This involved a predawn launch on Monday with a late-night return on Friday. A few times Glynn Lunney, who was an engineer working at one of the Lewis wind tunnels, flew with us on the Friday and Monday dead-head legs. He was courting his future wife, who at the time was a nurse at Langley. He later became a flight director and was a key leader in Mission Control during our Apollo 13 crisis.

I was assigned to be a copilot for Joe Algranti on my first aircraft ferry run to Washington and Langley. The day was pretty foggy. I looked out across the field and could not see the airline terminal or the Cadillac plant. I asked Joe if he had checked the weather and he jokingly answered, "Why check it? We are going anyway." In a little more than one year of these ferry flights, a flight had never been canceled for weather, and in that year I more than doubled all the actual instrument time I had logged before.

In the winter, we encountered conditions that would cause airline pilots to scream for altitude changes to avoid icing, but our R4D would just keep chugging along. The copilot's primary duty was to tackle the ice buildup on the wings. An inspection at the edge of the side window revealed the type of ice, from clear to rime, with clear ice being the worst. On the leading edge of the wing, there was a rubber boot that could be pneumatically inflated to periodically break off the ice. In order to retain visibility out of the front windscreen for landing, a small release of glycol did the trick.

Once, the snow and ice won the battle when I made a short trip down to Wright Patterson AFB. The weather forecast for Cleveland called for heavy snow through the night. When we were near Cleveland, approach control directed us to enter a holding stack of planes to await clearance for landing. We got a call from approach control that a Piedmont aircraft could not clear the piles of snow on the runway, and Cleveland Hopkins airport was closed. I requested a low-frequency approach to Burke Lakefront Airport just a few miles away, using the stadium ADF radio near the Cleveland Browns stadium. We broke out below the clouds at 400 feet. Rows of packed snow formed across the runway from the blowing wind. When I landed at Burke Lakefront, it felt like the plane had flat tires as we bumped over the snow.

In the spring of 1960, Jack Enders and I were assigned to start up a zero-G flight program. Our Lewis Research Center program would be the second in the US, following the earlier Air Force zero-G programs at Holloman AFB, utilizing an F-94C, and at Wright Patterson AFB utilizing a Convair C-131 transport aircraft. The Air Force program had the mission of conducting physiological testing on human subjects. Our Lewis zero-G program tested various fluid systems for rockets such as the Centaur and the SNAP 8 space nuclear power facility. Several aircraft were evaluated to assess whether their bomb-bay volume was sufficient to free float the test specimens. We chose the North American Navy AJ-2 Savage for its performance capability and availability, since it was being phased out by the Navy. The AJ-2 was a medium bomber designed to operate from an aircraft carrier and was unusual because it had three engines: two reciprocating Pratt & Whitney R-2800 engines with 2,300 horsepower and an Allison J33 turbojet. Its bomb bay was just five feet by five feet by thirteen feet.

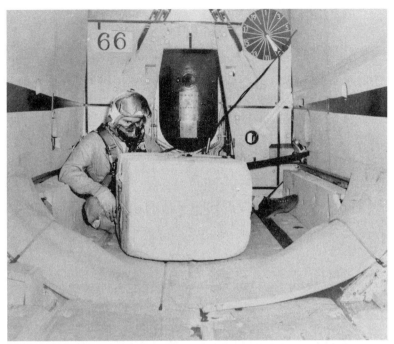

Conducting a zero-G experiment package while in the payload bay of the AJ-2. Courtesy of NASA

I flew my first familiarization flight in the AJ-2 on May 20, 1960. The Navy provided us with all of their AJ-2 unique avionics and five AJ-2 aircraft at the Navy Litchfield Park boneyard in Phoenix to scavenge for spare parts. We had meetings with Air Force zero-G project personnel at Wright Patterson AFB to garner any lessons learned. From their early experience, it was clear that we would need to modify our AJ-2 to assure that the reciprocating engines were not starved of an oil supply. Two special free piston reservoirs were installed, so that on each zero-G trajectory, there would be sufficient oil. The instrument panel was modified to allow the pilot to fly the trajectories as smoothly as possible. The key aids were the standard gyro horizon, acceleration readouts in three axes, and a TV monitor showing the action in the bomb bay.

Flying the maneuvers was tiring, so Jack Enders and I normally switched out during each flight. A trajectory would involve a dive to attain around 375 knots with a several g pull-up to approximately 30 degrees. As the speed bled off, one would go over the top at 240 knots to recover, in a 30-degree dive to a speed of about 390 knots. Depending on the pilot's control, between twenty-two and twenty-seven seconds of zero G were attainable.

The largest free-floating experiment flown involved liquid hydrogen. It weighed more than 150 pounds due to the heavy outer bombproof casing. An electric hoist was added to the payload bay to redeploy the heavy metal sphere to the center for the next zero-G trajectory. When acting as the bomb-bay attendant I hurriedly moved into position behind the protective wall at the start of each run, so I would not end up with a heavy sphere of liquid hydrogen sitting on me during the aircraft pullout. An excellent summary of the program is captured in the NASA Technical Note D-3380, "Use of Aircraft for Zero-gravity Environment," dated May 1966.

———

To be closer to my squadron at Mansfield Airport, my wife and I moved our family to a rental home in Ashland, Ohio—the fifth move of our marriage. Our second son, Stephen William, was born four months before the recall of the Ohio ANG 164th Spider Tactical Fighter Squadron. We now had children who were Texans, Okies, and Ohioans.

During the year on active duty, I enjoyed more than 300 hours in the F-84F. This included conventional weapons delivery sorties to practice bombing and strafing, as well as completing air-to-air missions at the Indiana Atterbury weapons range, or deploying to Volk Field in Camp Douglas, Wisconsin, and Hunter AFB in Savannah, Georgia. But, with the ever-increasing Russian threat, we shifted to special-weapons training to launch a nuclear weapon. Our planes were modified to include a T-270 control unit in the cockpit to set up and arm an atomic weapon. Specialized training included flying low-level sorties, below 500 feet, across West Virginia to execute the low altitude bombing system to deliver a practice bomb at the Atterbury Range in Indiana. The bomb was released near the vertical to allow time for the aircraft to escape while the bomb continued upward. If a real bomb was delivered, the cockpit mirrors would have been turned inward, and one would wear an eye patch to prevent exposure to the blinding flash. We practiced flight plans, assuming that we would launch from a French airfield to an assigned target city, in an Eastern Bloc country. Strategic Air Command bombers and nuclear subs were also ready for action, and most squadrons in Europe were on high alert. I was concerned about the situation escalating. In my opinion, the general public had no idea of the magnitude of the threat. Over the years, I have come to realize the insanity of humans having so many atomic weapons in so many countries, because unstable leaders and regimes could, theoretically, destroy the world as we know it overnight.

I left active duty in August 1962 and returned to the Pilot's Office at Lewis Research Center. There were some personnel changes in the office: Jack Enders had transferred to the Aeronautics Directorate at NASA headquarters and Clifford C. Crabbs, whom we called "C-Cubed," was his replacement. Cliff did not have a military background, but was active in civil aviation. He owned airplanes, including a Mooney Mite and a Cessna 190, and a helicopter. The Mooney was so small that Cliff could stand in the cockpit and reach over to manually turn the propeller to start the aircraft. I visited Cliff at his house and instantly recognized that he was a bachelor. There were piles of *Aviation Week* and *National Geographic* magazines everywhere. Most of the countertops and the mantles were covered with nuts, bolts, and airplane components. He kept the helicopter in a barn and had horses in two stables because he participated in harness racing. A three-legged dog roamed around.

Lewis had a second zero-G project involving the Aerojet Aerobee sounding-rocket that was launched from the NASA Wallops Station. These suborbital flights exceeded altitudes of two hundred miles, so there would be longer periods for zero-G experiments. A contractor had been selected to develop a parachute recovery system for the zero-G payload that rode on the Aerobee rocket, but this system was having problems, so the effort was taken over by NASA and successfully completed. Bill Swann was the lead pilot on the project.

One night, we received an urgent call to deliver spare pyros, or pyrotechnic devices, to support an Aerobee launch that was scheduled for the next morning. An added challenge was making the delivery to the airfield by NASA Wallops Station, which was closed at night. The pyros were safely packed in Styrofoam blocks and sealed with duct tape. Bill Swann and I set up a low slow pass in our NASA R4D for the package drop, aiming for the grassy area beside the main runway that was illuminated by the headlights of two trucks. The pyros were installed for a perfect Aerobee launch in the morning. I wonder how long the reviews with formal documentation would have taken to support that launch today.

During my time at Lewis, I made several trips to NASA Centers at NAS Moffett Field and Edwards AFB in California to deliver articles such as a prototype of a ramjet that would be shown at NASA's annual open house. This event updated the general public and media on the projects underway.

My career goal, at that point, was to transfer to the mecca of NASA flight research at Edwards. Finally, in November 1962, Joe Walker, chief pilot at Armstrong Research Center, let me know there was a position available. I was overjoyed: The Standard Form 50, Notification of Personnel Action, listed me as a GS-12 aerospace research pilot and engineer, with an annual salary of $9,475.

—

CHAPTER 5

THE X-SERIES

As I came around a bend on Highway 58 and saw Leuman Ridge and Rogers Dry Lake at Edwards AFB, I was on a high, thinking about how I would soon be airborne at this great flight test facility. This was on the family drive during our move to California, but when I turned to tell Mary how happy I was, the expression on her face made it clear that she had different thoughts about the desert scene before us, which was a far cry from the Mississippi Gulf Coast.

Our new rental home in Lancaster, California, was on a dusty street. The yard did not have a single blade of grass. Because I've never been a fan of mowing grass, this was not a problem for me, but we purchased a home at 44561 Leatherwood Avenue within six months where there was plenty of grass.

Lancaster is in Antelope Valley, which has a variety of plants that I had not seen before, including the impressive Joshua tree. In the spring, a beautiful crop of poppies bloom in the valley that is otherwise arid. Joshua Tree National Park was not far from where we lived, though Edwards AFB was quite a drive from our home. It was even farther for my friend and comrade, Don Mallick, who lived in Quartz Hill. Don and I would often carpool for the long drive. In the heat of summer with temperatures rising above 100 degrees Fahrenheit, we would make a "safety" stop at the Long Branch Saloon on Sierra Highway to

At Edwards AFB in front of a T-38A in 1964. Courtesy of NASA

ease our hydration problem with a pitcher of beer. I had no air conditioning in my auto and Don had a modified system that had questionable performance and integrity.

Edwards AFB was self-sufficient. It had everything from base housing and a community college, to a hobby shop and hospital. There were also many hangars because any number of aircraft test programs were going on at any given time. Most of the roads were named after test pilots who served at Edwards, including many who lost their lives during flight tests. The former Muroc AFB was renamed after Capt. Glen Edwards, who died in a crash of a Northrop YB-49 flying wing in 1948.

On my first day at Flight Research Center, known today as Armstrong Flight Research Center, I was greeted by an X-1 rocket ship mounted on a pedestal at the front of the building. I found my way to the Pilot's Office, on the second floor, and my coworker Bill Dana pointed out my desk. All the pilots had desks in one large room, except Joe Walker, who had a small office with glass windows at the end of the room. Joe Vensel, flight operations chief, was in an office on the other side of Joe. One secretary, Betty Callister, took care of the whole office. Just as at Lewis, I had a great view from the window looking out over the Rogers Dry Lake and down a long taxiway leading to the main Edwards AFB runways. The Army and Pratt & Whitney occupied the nearest hangars. Further along was a string of hangars with the last one housing the Aerospace Pilots School aircraft. NASA was at the northern end of hangar row.

My first flight at Edwards was with Jack McKay and Milt Thompson in NASA 136, a C-47. I was impressed to be with Jack, who had flown every one of the earlier X aircraft, and during this era he was flying the X-15. Milt was working on the early-concept phase for the joint NASA-Air Force Lifting Body Program and would later fly the X-15.

There were already thoughts of what might come next, including flights to Mars. The lifting body shape was thought to have the capacity to land on Earth, when returning from Mars. The various configurations evolved from wind tunnel model tests at NASA Langley and Ames Research Centers. The M2F1, affectionately called the "Flying Bathtub," was the first one built.

In what could be considered a precursor to my experience flying the *Enterprise* several years later, I started glider training at Tehachapi Airport, nestled within the surrounding mountains. The location was ideal for providing the lift needed for gliding. This training was deemed a prerequisite for those of us slated to be tow pilots. My instructor Fred Harris, with Skylark North, coached me through three flights in the Schweizer 2–22, after which I flew solo in the Schweizer 1–26. Most of the flights were launched from the air, towed aloft by a light plane. I was immediately struck by how quiet gliding was. There was no engine noise and I felt no vibrations at all. Passing over a school yard, I could hear the laughs and shrieks of children playing. Gliding was so enjoyable to me that I took my wife and daughter for flights.

———

Through 1963, I lived out a pilot's dream: I flew a variety of aircraft and was involved in a number of test programs almost every day. For example, Bill Dana and I flew a Navy A5A, in a simulated supersonic transport to Los Angeles International, in conjunction with the FAA, to evaluate air-traffic-control's handling of the supersonic airplanes that the British and French were developing.

Bruce Peterson, noted for his dramatic crash in a lifting body test that inspired the opening scene of the TV show *Six Million Dollar Man,* gave me my first checkout flight in the F-104 Starfighter. The flight included an afterburner acceleration, where I first experienced Mach 2, or moving twice as fast as the speed of sound.

Chuck Yeager, the celebrated World War II hero and first pilot to break the sound barrier, was scheduled to fly the M2F1 lifting body vehicle, and I was asked to take him on a chase flight in our Cessna T-37 Tweet aircraft to observe the operation. I looked at this as a great opportunity to display my flying skills. I had flown a number of chase missions with a NASA photographer, so I had a scheme worked out to stay with the lifting body, even during its steep descent to landing. After takeoff, I flew in formation out to the side as the C-47 tow aircraft slowly climbed to 10,000 feet with the M2F1 in tow. Approaching the release point, I shut off one engine on the T-37 Tweet and, at release, lowered the landing gear, deployed speed brakes, and shut off the second engine. Even with all the added drag and the fact that there were no engines, we would start to overrun the M2F1 in its 30+ degree dive. At this point, I executed my usual planned barrel roll around the M2F1, pulling 3 Gs to kill off airspeed. After two of these maneuvers, I followed the M2F1 to its landing on the Rogers Lakebed. Using the speed remaining, I rolled to the NASA ramp and stopped the T-37 Tweet in the exact spot that we had departed from. I snidely told Chuck that I did not restart an engine, because I did not want to stress the older lead-acetate batteries.

During this period, I logged an average of thirty-seven flights a month, which amounted to more than one flight a day. Many were in support of the X-15 program. These sorties included checks of the appropriate alternate lake beds. There were a series of lake beds we used for emergency landings, including Cuddeback, Three Sisters, Hidden Hills, Mud, Delamar, Smith, and Rogers (the launch and recovery lake bed). For the lake bed checks, the aircraft of choice was the C-47, because of its large tire footprint. The dry lakes could be deceiving because though appearing to be dry, they were sometimes wet. That was the case when Neil Armstrong and Chuck Yeager were checking out Mud Lake, Nevada, in a T-33A. Against Chuck's advice, Neil attempted a touch-and-go landing, but the aircraft's tires got stuck in the mud, which is where they were mired until a pickup came to the rescue. If the lake beds could talk, they would tell of a number of unplanned X-15 landings by Scott Crossfield at Rosamond, Jack McKay at Mud Lake, and Milt Thompson at Cuddeback.

Mary and I loved living in Lancaster. We often took the kids to Santa Monica Beach, Knott's Berry Farm, or Disneyland, which were all just two hours away. The Antelope Valley was scenic. We had family picnics in the surrounding Tehachapi Mountains, many with Don Mallick and his family. In the winter, we spent many a day enjoying the snow in the San Gabriel Mountains.

———

CHAPTER 6

MY INTRODUCTION
TO AEROSPACE

Don Mallick and I began the Aerospace Research Pilot School in 1964. Don and I were the only civilians among the sixteen students in Class 64A. There were two foreign military students and the remainder were either Air Force or Navy. It was a one-year program. The first six months included classroom lectures and flight tests that covered aircraft performance, flying qualities, systems, propulsion, stability and control, and test analysis. The second half included classroom training and flight missions oriented toward space flight. Chuck Yeager was the commandant of the school. The actual test flights were in selected aircraft that were chosen to analyze a handling quality or performance characteristic.

Three classmates and I planned and conducted a short test program focused on the afterburner climb performance of the T-38A aircraft. Each run began with a pull up at a designated airspeed to a planned climb angle. On one run, I started the pull up at 529 miles per hour and heard a loud noise, which was accompanied by an abrupt yawing motion. This was another *never panic early* moment, so I quickly reduced the throttle and was relieved that we were still headed in the right direction, namely upward. We surmised that a landing gear had been deployed at 253 miles per hour over its listed maximum deployment speed. Edwards's AFB tower was informed of the potential emergency and an aircraft was sent to look us over. The problem was that one main landing gear had been deployed, so we moved the landing-gear handle down and got three green lights in the cockpit, indicating that all of the landing gears were down and locked. I used a long, straight-in approach to make the landing as soft as possible because I was concerned that the landing gear that deployed too early would collapse. We rolled out to a complete stop straight ahead until the landing gear safety pins could be installed.

One Friday, Don and I decided to stay late to take care of some data analysis, rather than drive all the way to our homes in Quartz Hill and Lancaster, and have to come back to the base on Saturday. We called our wives to let them know we would be working late, but our plan fell apart when Chuck Yeager ran us out of the office to join the happy hour at the Officers Club.

——

June was the last month of the test pilot school portion of the program. Our class was scheduled for a trip to observe aircraft carrier operations. We flew to San Diego the day before and took in some of the action at a favorite Old Town watering hole that evening. The next morning, we were flown out to the ship in a Grumman carrier onboard delivery, or COD, aircraft. It was a great day to see the carrier pilot qualification landings in nearly every type of Navy aircraft. The fury of the noise and deck crew activity was impressive.

A graduation banquet was held at the Edwards Officers Club, and Don and I received our certificates of completion. I was awarded the A. B. Honts Trophy as an outstanding member of the class. All of the awards were named after a previous graduate who had died in the line of duty. Capt. A. B. Honts was killed testing a B-57 aircraft at Edwards on February 8, 1955.

I had already packed up my gear and moved back down the flight ramp to NASA when I received a call to meet with Chuck Yeager at his office. He offered me an opportunity to complete an additional part of the school, which focused on aerospace. Prior to this, only US military officers were permitted to take part in this training. I told him that I would have to check with my bosses at NASA, but he told me that he would take care of it.

I continued classroom study and flying for the next six months. The courses were oriented to space and included astronomy, bioastronautics, aerothermodynamics, and spacecraft design. Several simulators provided exposure to operating a spacecraft in a crew cabin and another simulator required controlling motion with attitude air jets. We flew in a converted KC-135 tanker aircraft for zero-G exposure. Similar to the later KC-135 that NASA used, the walls were padded and most of the passenger seats were removed to add maneuvering room. On each run, one could experience twenty-five seconds of zero G. Our class had a field trip to the Lockheed Martin Titan facility in Colorado that was constructing the Titan rocket. This was my first experience seeing a full-scale rocket up close. That was also the year of my first trip to Europe.

The traditional European trip to visit foreign flight test centers was scheduled for September. Chuck Yeager also planned a hunt in the Black Forest in Germany. A week before we were scheduled to leave, I visited him to let him know that since I wasn't officially in the class, I would return to NASA until the class returned. He responded that it was about time to tell the Pentagon that I was in the school and that I should make the trip. When I told him that I didn't have time to get a passport, he told me not to worry because he would get me an Air Force flight suit to wear and, since we would be flying on a USAF aircraft, no one would check the manifest. So off we went, with a stop at McGuire AFB in New Jersey, and the long trip across the Atlantic to Frankfurt, Germany.

Chuck Yeager was right—most customs agents only glanced at the manifest, so my not having a passport was not an issue. Chuck had been a squadron commander at Bitburg AFB. During that time, he befriended a Black Forest game warden. Soon after we landed, Chuck left the group to go on his hunting expedition. We continued our tour to Farnborough, England, to visit the Empire Test Pilots' School for briefings. We stopped in a nearby small town and visited a typical English pub, where some of the locals played darts, just like I had seen in the movies.

Our next stopover was at the Italian flight test center, Pratica di Mare Air Base, not far from Rome, Italy. I had the opportunity to fly the G-91 aircraft on a forty-minute flight with Georgo, an Italian test pilot. The G-91 was a fighter-bomber built by Fiat Aviazione, which later merged with Aeritalia. It looked like a smaller version of the F-86 Sabre jet. After Georgo started the engines and made a quick systems check, he called for taxi. He spoke to the tower in loud, rapid-fire Italian. I almost thought we had an emergency because his tenor was similar to what I would use to call a Mayday. Georgo demonstrated the G-91's maneuvering capability, including its ability to fly inverted for some length of time. After the flight, we were treated to a luncheon of many courses at an impressive mansion on a hilltop overlooking Sophia Loren's residence. Unfortunately, even though we peered over the edge to get a glimpse of her, we didn't see her.

Our final leg of the journey was to the Étain-Rouvres Air Base, for schedule briefings at Supreme Headquarters Allied Powers Europe. A few days of free time were scheduled for us to tour Paris. When we arrived at base operations, a French gendarme began to meticulously go through the manifest, checking credentials. Where was Chuck Yeager when I needed him? He assured me that I would not need a passport. I stayed at the end of the line as each of my classmates showed their active-duty IDs or passports. I handed the gendarme my red military-reserve ID and he shook his head from side to side to signify *no*. My compatriots stood on the other side of the room near the exit, snickering and thinking that I was going to spend the next couple of days in base operations, and they were going to have fun in Gay Paree. I pulled out my NASA pilots license, my FAA pilots license, and my driver's license. After each one, the gendarme shook his head. As I began to resign myself to buying my meals from the candy and soft drink machines for the next few days, the gendarme spoke in French to the staff sergeant standing by. He told me that, as best as he could interpret it, the gendarme said, "I cannot stand to see a grown man cry. He can go to Paris if he promises not to misbehave!" Had it not been for the fact that French-US relations were falling apart, I am certain that I would have been routinely waved through.

We returned to Edwards for more test flights—this time in a few 100 series aircraft. The F-101N variable-stability aircraft, like the F-104, had a high horizontal stabilizer and had to be controlled to stay below a critical angle of attack to avoid becoming unwieldy. I also flew seven flights in the F-106 Delta Dart to assess stability and control, and dynamics. Several flights occurred at high Mach and high altitudes up to 50,000 plus, which demonstrated its intercept capability.

At the NASA Flight Research Center in front of the F-104N in 1966. Courtesy of NASA

We visited the School of Aviation Medicine at Brooks AFB, in San Antonio, Texas, that had a centrifuge to provide us with runs up to 14 Gs. A primary lesson learned was to move slowly and not turn one's head rapidly after exiting the rig. This prevented a spinning sensation that could cause vomiting. In addition to lectures, we also witnessed the effect of altitude sickness that led to dog blackouts. Another highlight was observing a human autopsy. For some reason, it was another part of the planned activities laid out by the School of Aerospace Medicine.

The Manned Orbiting Laboratory Program was highly classified, and I did not know that a selection process was ongoing, until it was announced that our two US Navy members of Class 64A—Dick Truly and Jack Finley—were chosen to join this initial group of astronauts. The program coupled a Gemini capsule with a laboratory vehicle to perform reconnaissance missions in orbit.

In December, we flew a number of what were called "Zoom" flights in the F-104A aircraft. Runs were made down the supersonic corridor at 35,000 feet and with max afterburner to achieve Mach 2. On succeeding flights, incremental pull ups were made to a maximum of 45 degrees pitch angle that could achieve an altitude of 90,000 feet. These flights provided experience wearing the USAF full-pressure suits. The flights also provided the pilot a rudimentary X-15 altitude profile flight. One had to monitor the exhaust-gas-temperature gauge on the climb to make sure to shut off the afterburner when the gauge reached the red line, and again shutting off the engine when getting to thinner air. Careful control was required over the top to keep from exceeding the critical angle of attack and going into a spin. A Navy lieutenant in the class before us was fresh in our minds, because he spun out and could not recover. He was forced to eject and the aircraft was lost. In a normal recovery, the engine would be restarted for a landing back at Edwards.

—

Back at Flight Research Center, my first test assignment was with the General Aviation evaluation program. The Civil Aeronautics Board requested that NASA evaluate seven different airplanes because there was an increasing accident rate in general aviation civil aircraft. I took Don Mallick's place because he was assigned to test the lunar landing research vehicle, which was the predecessor of the lunar landing training vehicle. Later in my career, I flew that craft twenty-two times to train for the Apollo 16 Mission.

The general aviation evaluation program was set up to perform full stability and control evaluations, including oscillograph data collection, on five of the seven aircraft that included a Beech S-35 Bonanza. Qualitative evaluations were performed on two of the aircraft using hand-and-foot force measuring tools. Toward the end of the data collection for each aircraft, a series of evaluations were conducted with pilots that had varying degrees of flight experience. General aviation company test pilots also took part in these evaluations, but they were not permitted to evaluate their own company's aircraft. I was a coauthor of NASA Technical Note TN D-3726, *An Evaluation of the Handling Qualities of Seven General-Aviation Aircraft,* which covered the program. I also delivered a paper on the test program at the tenth Symposium of the Society of Experimental Test Pilots and won the Ray E. Tenhoff Award, which was named after the Convair test pilot and first president of the Society of Experimental Test Pilots who was killed in a B-58 Hustler accident.

Up until the NASA announcement for the next astronaut selection on September 10, with a requirement to file for selection by December 31, 1965, I continued to enjoy flying. I completed tests in the Piper Cherokee 150, Cessna 310, Aero Commander 680F, Cessna 210, and Piper Apache. While at Cessna to pick up the 310, I flew an evaluation flight in the Cessna 411. Across those first eight months of 1965, I had also flown thirty-four support flights in the X-15 program.

I also participated in several other test programs. I flew nine sorties in our T-33A in order to evaluate the minimum field of view required to perform an unpowered approach. This test entailed applying orange material to a portion of the canopy while wearing blue goggles to assess the varying degrees of visibility out of the window. This concern arose during the X-15 experience, where windows fractured on several flights from heat built up. These flights were conducted at North Base, to be clear of Edwards's normal traffic. I flew fifteen flights in the Calspan variable stability NT-33N aircraft, verifying cases describing the handling qualities for the upcoming M2F2 lifting body that was being built at Northrop. Milt Thompson and Bruce Peterson flew evaluation and training flights for their upcoming M2F2 flights, launching from the B-52 aircraft.

———

Astronaut selection had progressed at NASA. In 1959, Group One, as the original seven Mercury astronauts were known, was chosen. In 1962, the New Nine were added to help get us to the Moon. Fourteen more were chosen in 1963 and, in 1965, six scientists were added. One day, Neil Armstrong visited the Flight Research Center after leaving that assignment for the Astronaut

Office. My friend Don Mallick asked him what it was like being an astronaut. Neil's answer was, "It's a lot of meetings, with a lot of time sitting in simulators, and not much good flying."

I evaluated my situation. If I stayed, I would likely never get to fly the X-15. Those assignments were made by seniority, and there were three pilots ahead of me. At that point, my best future opportunity was to fly in the lifting-body program. Over the previous nine months I was involved in six test programs, flying fourteen different types of aircraft. This was good flying, but the thought of going to the Moon—an unbelievably great adventure—was irresistible, so I submitted my application.

In my remaining six months at Edwards AFB before the selection announcement on April 4, 1966, I completed the General Aviation program. This included tests of the Piper Apache, Beech S-35 Bonanza, and the PA-30 Twin Comanche. Just before wrapping up testing on the PA-30, the chief pilot at the Los Angeles branch of the FAA requested that we conduct tests for flutter in the PA-30 because there had been several fatal accidents in the aircraft. One witness observed the tail breaking free.

All the aircraft we tested had manual control of the control surfaces through a pulley arrangement. The horizontal stabilizer was normally a two-piece control surface with the front part fixed to the fuselage and the rear portion movable to provide pitch and trim control. However, the PA-30 was configured with a horizontal stabilizer as a single movable structure. Many aircraft that I have flown had that configuration, but were hydraulically supported—either directly or as boost to reduce control forces for the pilot. In preparation for the tests, the PA-30 was modified to include strain gauges on the horizontal stabilizer that was hooked up to an oscillograph. Two motion cameras, set to run at twenty-four and seventy-five frames per second, were mounted in the cabin, facing aft toward the tail stabilizer. On each of the flights, John Manke flew chase with the Cessna 310 and a NASA photographer, who was using a high-speed motion camera running at four hundred frames per second. Russ Barber, a test engineer, normally flew with me on the data-collection flights, but because of the possible risk of flutter, I flew alone and wore a parachute.

The test approach was to trim at a speed well below the aircraft redline. The redline on the airspeed indicator was 205 miles per hour. Certification of the PA-30, under FAA regulations, allowed a higher speed, up to 239 miles per hour. Our first test was set at a conservative 180 miles per hour, with the intent to increase the speed by five miles per hour on each flight. After each trip, I landed to allow Eldon Kordes, chief of the dynamics and control division, and Russ to extract the data from the oscillograph for analysis. Of key interest was the zeta stability factor, with zero indicating instability.

After trimming the aircraft to 195 miles per hour, I struck the control yoke with my hand to excite vehicle dynamics. This was a *never panic early* moment. I felt like I was riding a jack hammer—the plane was shaking so hard that my vision blurred. Within a few seconds, the vibrations stopped. Fortunately, though the stabilizer suffered internal structural damage, it stayed attached to the plane. I slowed down and called Edwards's tower for a landing on the lake bed. There are great shots of the horizontal stabilizer twisting and moving up and down at NASA media and on YouTube.

———

On February 8, 1966, I reported to the Aerospace Medical Division at Brooke AFB in San Antonio, Texas, for a physical. The other candidates were twelve civilians, twelve men from the Air Force, sixteen Navy men, and sixteen Marines. The only one who I knew was Daryl Greenamyer, a Lockheed test pilot. The overall physical was thorough, and the only unusual part for me was the experience of going through several psychological tests and interviews by a shrink. A few weeks later in Houston for more testing and an interview, a NASA security officer escorted fellow candidate Hugh Jackson and me through the Rice Hotel's rear entrance to register for our rooms. I wondered what all of this spy stuff was about: How did the CIA get mixed up in this process? Hugh and I spoke briefly with NASA greeters Jack Cairl, Deke Slayton, and Warren North. We were informed that we were in—a formal selection announcement would be made on April 1, and we were to report for duty on May 1.

We spent the next day learning the definition of thirty space-related terms. We also wrote essay responses to questions that explored our knowledge about astronomy, space programs, and other areas. On our last day we toured the MSC that included Mission Control, a docking simulator, a centrifuge, a lunar landing research vehicle trainer, and the Gemini Mission simulator. The final step in the selection process was to appear before a review board chaired by Deke Slayton, with members Warren North, John Young, Mike Collins, and C. C. Williams. A number of technical questions were discussed, including Chuck Yeager's accident on a Zoom flight in the rocket powered F-104.

———

CHAPTER 7

LIFE ON THE EDGE OF SPACE

I officially transferred to NASA's JSC on April 24, 1966. I hitched a ride to Houston with my new classmate, Joe Engle, in his Mercedes. We both left our families behind—they would make the move later. In my case, Mary and the children would come after we sold our home in Lancaster. My job title was aeronautical research pilot and engineer. I was a GS-14, making an annual salary of $14,680. The Civil Service Registry did not list "astronaut" as a position title.

John Young dubbed us the Original 19. Among us, seven were from the Air Force: Charlie Duke, Joe Engle, Ed Givens, Jim Irwin, Bill Pogue, Stu Roosa, and Al Worden. The Navy members were John Bull, Ron Evans, T. K. "Ken" Mattingly, Bruce McCandless, Ed Mitchell, and Paul Weitz. The Marines were Jerry Carr and Jack Lousma. Four civilians completed the group—Vance Brand, Don Lind, Jack Swigert, and myself.

Two of us had combat tours in Vietnam, ten had been through test pilot school, and all of us had been through military flight training. I felt at home with my new comrades because we had similar backgrounds. I shared an office with Ed Givens.

Ed Mitchell, Ken Mattingly, and I camped out in a three-bedroom apartment in Dickinson, Texas. For transportation, I bought a secondhand VW Bug—the first of several of those reliable vehicles that I owned—because of low operating costs. Ed, Ken, and I shared a deep friendship through challenging, turbulent times during our future missions.

Ken was intense and one of the most detail-oriented individuals I have ever known—an important trait to staying alive in this business. He seemed to sleep only five hours a night and feasted during the day on cheese, peanut butter, and crackers when at the apartment. Ed and I were straight and narrow types. We left our abode within the first two months, when our families made the move to Houston.

The Original 19 studied astronomy, aerodynamics, rocket propulsion, communication, space medicine, meteorology, navigation, orbital mechanics, and geology. For me, much of the material was a refresher, except for geology—that was an entirely new experience. To acquaint us with what the professional geologists and scientists might do with samples that we would bring back from the Moon, we had an introductory geology lab where we studied thin sections of rock samples and discussed their chemical makeup.

On June 10, Joe Engle and I scheduled a flight to Edwards to attend funeral services for Joe Walker, who had been my boss at NASA Flight Research Center. He held the altitude record for powered aircraft in the X-15 for years, exceeding the Karman line at one hundred kilometers twice. Joe was in a NASA F-104N, with the B-70 and several other aircraft in formation for a photo shoot. A midair collision occurred resulting in the death of Joe and Maj. Carl Cross in the B-70.

In preparation for our being the face of NASA at public events, we were trained by NASA's Public Affairs Department on how to engage with an audience. I worked with a couple of individuals, including Gene Marianetti, who served as our protocol officer during the post-Apollo 13 travels. Gene prepared me for what we called "our week in the barrel," which was a weeklong series of events—often several a day. I spoke at schools, luncheons, banquets, and any other events that needed a speaker. I had a repository of general NASA program information and spacecraft details stored in my brain to draw from. I never knew what my week in the barrel would bring, but I learned to talk from what I knew, rather than just reading from a prepared speech that was given to me by a professional speechwriter.

Our first geology training in the field took place at the Grand Canyon. Each trip was led by an instructor who had extensive knowledge on the area being visited and could cite major features. The various rock formations gave me an instant perception of geological time, versus the concept of great distance that astronomy offered. The present day layer that I saw at the rim of the Grand Canyon, all the way to the bottom layer on the canyon floor, represented more than a billion years. It was hard to conceive that the two large limestone beds we were clambering around on had once been under large inland seas. The thick Coconino Sandstone layer were remnants of the sand dunes that rival the Sahara Desert and extend across four states.

Other field trips exposed us to fumaroles and ash flow at the Valley of Ten Thousand Smokes within Katmai National Park in Alaska. One of the volcanos in the area emitted a small amount of white smoke. We also visited a lake adjacent to a forest ranger camp. While there, we were offered an opportunity to fish. The ranger warned us to whistle or sing while on our short hike, to make sure that the Kodiak bears knew of our presence. Charlie Duke, Paul Weitz, and I donned waders for the cool water to fish for trout, but the dying salmon, near the end of their spawning run upriver, got in the way with inadvertent hooking. Charlie and I were on one side of the river and Paul was on the other, when we saw a huge Kodiak bear part the branches behind Paul and scoop a huge salmon out of the river to eat like a lollipop. Following the ranger's advice, Paul bellowed out in his deep voice, "Hey bear!" The bear reared up and let out a ferocious roar. Paul literally walked on water and made it to our side of the river, even with his waders full of cold water. He reasoned that the bear could only attack one person at a time, so he crossed the river to join us. The Original 19 almost became the Original 18.

We then went to Iceland, which was the first place I witnessed the sun not setting fully, leaving the sky twilight at midnight. In addition to viewing other volcanos and igneous rocks, we saw a glacier, which was an incredible sight.

On the big island of Hawaii, many of our survey and sampling exercises were done in the saddle between Mauna Loa and Mauna Kea. Since we were staying at a small Army camp near the volcanos, Jim Irwin convinced the Army to open their theater for one night of entertainment. We suffered through a private showing of the Italian film, *Revenge of the Gladiators*. Some of us visited a prison camp on the Hilo side of the Big Island, which was arranged by local officials. We gave a brief talk about our Apollo Mission and why we were training in Hawaii. For that, we each received a set of Monkey Pod wooden salad bowls that I enjoyed for many years.

The Original 19 went on ten geology field trips to sites that included volcanic areas such as Bend, Oregon. That's where we witnessed an obsidian flow, which is made when lava cools so rapidly that the formation resembles glass, as opposed to a crystal structure that results from slower-moving lava. We had to wear gloves and be very careful climbing around on that terrain so as not to cut ourselves. Other exercises took place at Valles Caldera near Los Alamos, Arizona, and cinder-cone areas in Flagstaff, Arizona, and Medicine Lake, California.

It never failed on our early trips—someone would notice that one of our group was MIA. It was always Bruce McCandless. He was an avid bird watcher and would stray from the group to observe a bird of interest, so Ed decided to appoint a "Bruce Watcher" on each of our subsequent field trips. It is ironic that the first person NASA allowed to be completely free, untethered, and on his own in space, with a manned maneuvering unit strapped to his back, was Bruce. Other trips to major facilities that were part of the Apollo program took us to Marshall Space Flight Center (MSFC) and Kennedy Space Center (KSC). We also visited major contractor facilities at North American in California and Grumman on Long Island in New York.

We were given several days of briefings about the command and service module (CSM) and LM that all of us hoped to fly. NASA educators created a viewgraph show that featured simple schematics of the various systems in the spacecraft.

Because of a potential abort from orbit, we received survival training in the desert area of Washington state and jungle survival in Panama. We were also trained to survive in water. Desert survival training was the worst, because it involved sitting around most of the day while minimizing conversation, because talking causes a loss of moisture. Most lessons about shelters and creating devices in the ground to capture dew moisture took place after dark.

During desert survival training, I wrapped myself in parachute material as protection from the hot sun. Courtesy of NASA

In each exercise, we were grouped in threes, just as we would be in a capsule crew. Water survival involved, as it had in my pilot training, the Dilbert Dunker at NAS Pensacola. It also entailed deploying a small raft and learning to operate the signaling equipment. It was just my luck to be teamed up with John Bull and our group wanderer, Bruce McCandless, for jungle survival. Our training was run by the USAF at Howard AFB, with the support of Panamanian Guards. We flew by helicopters to our respective camp sites. Once there, the items that would be found in the capsule, plus one section of a parachute, which was used for hammocks to sleep in, were at our disposal. There is no shortage of water in the deep jungle. It rained almost all day and night, with only a few hours free of rain at night. Everything stayed wet. Starting a fire with the small book of paper matches or the flint and cotton provided in the survival kit was impossible. A guard came by once a day to check on our condition, and when we complained about the wood being too wet for a fire, he offered to help: After piling up some of the wet wood, he pulled out a flare to ignite and stuck it under the wood. Pretty soon, water bubbled out of the wood and we had a fire. From then on, our crew of three had to add fresh wood around the clock.

Naively, I expected there to be an abundance of animal life that we might capture with a snare or kill with our machete, but I quickly realized that most animals lived in the overhead vegetation that obscured the sky. We could hear them above us. We ate hearts of palm that we cut from small trees. By the third day, we enjoyed a stew of bugs and small minnow-size fish. A survival rule is that if you boil things long enough, you don't have to worry about bacteria.

Of course Bruce strayed off, seeking birds and other critters. At dusk, John wanted to go hunt for our fellow crewman, but I convinced him that we would soon all be lost trying to find Bruce. I thought the best solution was to blow the whistle from our survival kit every fifteen minutes, so Bruce would be able to find his way back to our encampment. He eventually showed up empty handed.

After five days, the guard came to lead us out. He took us down to a river to the small raft that was included in the capsule-survival equipment. We were to ride the raft to a village down the river, but on the walk down the trail to the water, Bruce spotted a snake. Not just any snake, but a fer-de-lance, a highly venomous pit viper. Bruce thought it would be great to capture it and take it back to present to the Houston Zoo. Who else but Bruce would know that this snake was one type that the zoo did not possess? This was the most dangerous snake in Central America, and without prompt treatment after a bite, one could die. One of the Panamanian guards and a NASA doctor caught it by putting a gunny sack over its head. I think the doctor volunteered because he worried that the loss of an astronaut on his watch would not look good on his record, but Bruce insisted that he personally take it down the river. Others from our group looked at us smiling as we got ready to board our raft. They saw this as an opportunity for the Original 19, who were competing for a mission, to be significantly reduced by the bite of a snake. John Bull and I decided to ride outside, holding on to rope attached to the raft. We even faced a small set of rapids, but preferred splashing through them over being bitten by a poisonous snake. I paid close attention, just in case the raft capsized—I wanted to clear the way for Bruce and his "little pet." The snake made it to Houston and Bruce happily presented it to the zoo.

At the village, we communicated with rudimentary hand signals and were served some of the indigenous people's delicacies—iguana, boa constrictor, and a brown field rodent. The latter, which could have been a cousin of our Nutria in the US, tasted the best. In a show of appreciation to the chief for his tribe's support, we invited him to see a Saturn V launch several years later. The chief was impressed by the noise and fire, but was more impressed when he rode in an elevator at NASA's headquarters in Washington. He was amazed to see the doors close, then open up at a different scene, as if by magic. It seemed that this was his "Beam me up, Scotty" moment.

—

Six months had passed and our rookie training was over. Alan Shepard, chief of the Astronaut Office and America's first man in space, released a memo detailing our assignments. I was happy to know that I would be reporting to the LM/Lunar Landing Research Vehicle/Lunar Landing Research Facility branch headed by Neil Armstrong. It was encouraging to be part of the LM Development Program, because it was the vehicle that would land on the Moon. I was further encouraged to find out that I was one of ten members of the Original 19 who was chosen to receive helicopter training. The training was a short course conducted at NAS Ellison Field in Pensacola, Florida. Normally, the Navy helicopter training course was one year, but I flew an abbreviated twenty-one flights in the Bell TH-13M over eleven days with Marine Corps instructor Capt. Muyskens. I flew a little time in the NASA Flight Research Center Bell 13 with Don Mallick and Bruce Peterson coaching, where I had to learn to accommodate the relatively light control forces of the helicopter compared with any normal airplane. I also had to adjust to the requirement for left rudder input when increasing power to offset torque versus right rudder in a reciprocating engine winged aircraft. But in that short amount of time and flight hours, I had mastered hovering, auto rotations, and got to enjoy four solo flights. In addition to checkout flights in the NASA OH-13H at Ellington Field with other NASA pilots, I continued with helicopter proficiency flights throughout the Apollo program.

In December, crew assignments for Apollo 1 and Apollo 2 were announced. Ed Mitchell and I were assigned to Apollo 2 as support crew. Jim McDivitt was our commander for LM testing support. Ed and I reported to Jim's office to get our marching orders. We were raring to go. He simply told us, "I want you to go to Grumman and make sure that I have a good LM to fly!" Dave Ballard, an excellent NASA system engineer, worked with Ed and me over many days and nights ahead at Grumman. He did a tremendous job. His understanding of the vehicle problems resulted in outstanding follow-up solutions, and he coauthored most of our status report memos to Jim McDivitt.

Ed and I made our first cross-country flight to check out the Grumman LM Operations on October 9, 1966. There were many places to land in New York in order to access the Grumman plant on Long Island. Despite the fact that the runway was only 6,000 feet long—marginal for the T-38A aircraft, but suitable for the T-33A—I chose the airfield at Bethpage, because it was nestled within the primary Grumman Plant. From where I parked the aircraft, it was walking distance to Plant 3 or 25, where most of our time was spent in LM preparation. Ed landed at Calverton most of the time. On the other hand, Ed chose the T-38A, which had a shorter flight time, but then he had to arrange an auto ride to Bethpage from the Calverton airfield.

Grumman arranged a trailer for us to use that was a short walk to the Plant 3 high-bay clean room. The high bay is a room that is approximately three stories tall. The clean room helped prevent contamination of the vehicle electronics, as well as to avoid the deposit of debris in zero G. At the high-bay entry, we walked across a sticky pad on the floor that removed dirt from our shoes. Then we passed through a blower to remove dust particles, after which we donned a white coverall, white shoe coverings, white gloves, and a white head covering.

Our trailer had a small living area with bunk beds, an office area complete with a radio and telephones to monitor test operations, a television, and a small meeting area. This accommodation came in handy because test operations went on twenty-four hours a day, seven days a week. In 1967, test operations shut down completely only one day of the year—that was Christmas Day. Even though we normally had a room at a local motel, many times we would end up spending day and night in the trailer. It was convenient because we could hear the status of tests or climb into the bunk for a nap while awaiting a test start delayed for several hours.

The first day I entered the cavernous Plant 5 high-bay area it reminded me of when *The Wizard of Oz* changes from black and white to color. The scene in Plant 5 was just that vivid. People scurried around the facility dressed in white, clean-room garments, with an urgency—there was a schedule to meet so the US would be the first nation to land on the Moon. Multistory test stands lined one wall with several LM vehicles. LM-1 and LM-2 looked complete, with both ascent and descent stages. Another LM, the LM Test Article (LTA-1), was enshrouded with a special material to undergo electromagnetic interference testing.

———

Alongside its high hopes for a Moon landing, NASA was concerned that the continued LM delays were caused by Grumman's priority and emphasis to meet the concurrent Navy Aircraft program delivery schedules. NASA's top management and Lew Evans, the organization's president, recognized that Grumman had a "cultural" problem with NASA. Ed and I noticed this on the floor as well in our day-to-day dealings with Grumman employees, who were slow to accept the NASA way that seemed too bureaucratic.

I was in the trailer next to Plant 5 when I received a call on January 27, 1967, from Jim McDivitt, informing me of a fire on the launch pad and the deaths of the three Apollo 1 crew members. We lost Ed White, Gus Grissom, and Roger Chaffee. I attended Ed's funeral at West Point rather than the one at Arlington for Gus and Roger, because my daughter Mary Margaret was close friends in school with Ed's daughter, Bonnie. This accident, beyond the grief of the families, also had a major impact on the program. The investigation of the

LM-7 *Aquarius* at the Grumman plant getting ready to go to KSC and its Apollo 13 adventure. Courtesy of Northrop Grumman

incident uncovered the cause of the fire and led to a major redesign of both the CSMs, as well as the LM. It also led to major changes in management at NASA and its prime contractors. Joe Shea, Apollo program manager, shared blame for the accident, and was replaced by George Low. Soon after, Bill Lee, NASA's LM project manager, was replaced by Gen. "Rip" Bolender. For more than a month, there was no NASA top management at Grumman. During this period, Ed or I attended the meetings, assuming the role of NASA management.

Because of his vast experience and expertise, Tom Kelly was reassigned at Grumman on February 7, 1967. Since the schedule was slipping, and each day counted in the Space Race, Tom became head of LM Test Operations—also known as S/CAT or Spacecraft Engineering and Test. He is remembered as the father of the lunar module. From its conception through the finished design, he was the chief engineer. His book, *Moon Lander*, tells the fascinating story of designing that unique vehicle and should be required reading for any budding system engineer. Tom brought with him Howard Wright, another great engineer who had worked in the development of the A6 Intruder and the E2C Hawkeye. Both aircraft embodied systems that were the cutting edge of the digital world, integrating software and radar.

I think Tom quickly realized that this new job, in the world of nuts-and-bolts testing of his vehicle, offered quite a different challenge from his prior experience, where he worked with some of the best in the industry who had experience with Grumman. Conversely, the LM Test Operations team was a mix of long-term Grumman employees and many newly employed workers. In order to meet the demands of LM manufacturing and testing, Grumman grew by at least 7,500 employees.

Tom and Howard moved to temporary trailers next to Plant 5, which was known as the "Nerve Center." It was imperative for testing to be conducted the NASA way. These ritualistic methods, replete with detailed documentation, were developed through previous experiences at McDonnell during the Mercury and Gemini Programs. This level of detail and intensity was foreign to Grumman's experience developing and testing Navy aircraft.

The LM had systems that were, for the most part, stand alone. This was before the advent of electronic chips. The Apollo computer in the command module and the LM had approximately one-tenth of a megabyte of memory and was hand wired. The computer contained only capacity for guidance, navigation, and control, and had no connection to the environmental, communication, electrical, propulsion, lighting, or other systems aboard—they were all manually controlled by the crew with hard-wired connectivity. Later spacecraft and airplanes had many systems integrated through a central processor.

Depending on the vehicle configuration, a test plan involved seventy or more operational checkout procedures (OCPs). When all systems checked out, NASA approved the spacecraft's shipment to the KSC for launch readiness. Grumman assembled a group of consulting pilots that included supervisor Scotty McLeod and several cabin system operators to ensure that one member of this group would be in the cockpit if the power was on in a given LM. Scotty's team had basic vehicle-system knowledge and could protect against anything going wrong in the cabin that could potentially cause damage. This overlap of testing with a "hot" vehicle continued in the LM-5 buildup.

The test readiness review was a key ingredient of starting any OCP on the right foot. The test director led the meeting, offering a detailed description of the OCP and what was to be achieved. Meeting attendees included Grumman, NASA, and Resident Apollo Spacecraft Project Office (RASPO) personnel. Ed and I quickly found out that this meeting was a sham, often did not start on time, and the test director did not even set the stage for the test—not to mention that a number of key personnel were absent. The test readiness review was designed similarly to KSC's Mission Readiness Reviews that assessed whether the team and the spacecraft were ready for launch. Some of the absentees were NASA system engineers or quality control representatives, so I arranged a meeting with key RASPO management to let them know that, even though the vehicle was not ready to go through the rigors of the OCPs, NASA should be sure that they were not blamed for this.

I started a diary on April 11, 1967, that I kept from the time of my LM testing assignment to when I served on the Apollo 8 backup crew, through January 10, 1969. The first entry says:

Upon arrival at GAEC [Grumman Aircraft and Engineering Corporation] I was requested by John Johansen to sit on a higher level Board, surveying wiring Crabbs on LM-1. It appears that a special team comprised of KSC, MSC, RASPO have gone over the vehicle with a fine-tooth comb and written some 1100+ Crabbs.

A Crabb is what would have been considered a minor discrepancy, but following the Apollo 1 tragedy, new specifications were established for wiring. There were many problems: The wiring bundle ties were too far apart and there was too much slack in them; the bundle clamps were egg shaped rather than round; and there was discoloration of the insulation where a splice had been made. Some of the wire bundles sat against the sharp edge of a bracket. At the time, I considered this attention to detail excessive, but, several years later, I had a different perspective. Had it not been for this level of diligence concerning the wiring bundles, along with the added requirement of thoroughly potting the electrical-connector interfaces, we would likely have not survived Apollo 13. An electrical short was prevented, despite our diminished power and the buildup of moisture that was visible on our control panel and wiring bundles. Major wiring issues with LM-1 and LM-2 jeopardized their schedules, so NASA leadership decided to use LM-3 as the first manned LM.

———

Above the high bay area, from Automated Checkout Equipment Rooms with large windows overlooking the LMs below, the test conductor supervised the required procedures of the OCPs by radio. He had access to all participants in the ongoing test. Rather than glamorous or exciting, I found spacecraft testing to be somewhat ritualistic and agonizingly slow. For example, the test conductor read aloud each step in the OCP which was then tediously repeated by the designated tester to verify completion of the task. The test conductor commanded, "CDR, depress ABORT push button." Upon completion of the action, the tester responded, "CDR depressed the ABORT push button."

Quality control personnel were always present, but some steps in the procedure were flagged "Q," for quality points that required an actual physical stamp in the master OCP document that the test conductor kept. Some actions or data readouts that were deemed highly important required two stamps before proceeding—one by Grumman Quality and one by NASA Quality. There were many stops and starts. Until every step in a procedure was formally completed and signed off, the OCP was not closed. This could not happen until all the flight elements pertinent to the test were installed with every functional path verified end to end.

If anything happened, such as an unexpected data reading, a warning light, or if there was a missing expected action at any step, the test conductor called a hold. The situation was literally treated like a crime scene from the TV program *CSI*. In these cases, the test conductor created an interim-discrepancy document. Sometimes, when something went wrong in the testing process, when I was in the craft, depending on the length of the wait time, I would either lie on the metal floor to take a catnap or I would go back to the trailer for a longer rest. Over a year and a half, I had probably spent several weeks sleeping on the metal floor.

This testing activity went on around the clock. During all of 1967, test operations went on twenty-four hours a day, seven days a week, for all but two days that year—one was a snow day, the other was Christmas. Tests started at various times and could run for any length. For example, the first OCPs for the rendezvous and landing radars ran for twenty-three and twenty-seven hours, respectively. I was either in the vehicle or in the trailer during lengthy holds.

Ed and I normally checked into a nearby Holiday Inn, but there were nights when neither of us got back to our hotel room. This workload and pressure was hard on family life and was reflected in workers' reactions at times. One day, I went to the Nerve Center for a meeting that Red Haftner scheduled. When we got there, the room was occupied. Red slammed the door open with such force that he broke its upper window. As he did this he shouted, "OUT," from the top of his lungs. Another day I was eating lunch in the Plant 3 cafeteria when a fight broke out over the Arab-Israeli War.

I experienced heated arguments in meetings held in the astronaut trailer for the finalization of OCP. Early on, Dave Ballard and I were heavily involved in the development of a fully powered LM electromagnetic induction test, and we ran into resistance from some of the engineering designers who looked at it as an engineering test, rather than a validation of system performance. Our meeting, as in several to follow, ended up in an argument between Jimmy McCue, Fred Pullo, and a man named Toth from the LM-3 test team. Dave Ballard, Dan Mangieri, and I were there for NASA and tried to referee the disagreement that began when the test team primarily wanted to delete some of the redundant sequences in the OCPs. Dave and I wanted a set of steps added to the whole process that the cabin crew would execute at the test conductor's command to start without his further guidance. Since this was a first, there was some concern from the test team, but I assured them that this protocol would be supported by astronauts and they would be trained to ensure that there would be no operator errors. It was important to simulate the varied systems signals and mechanical activities that would occur during a lunar landing. The arguments could get very hot, so I used a police whistle to signal that everyone needed to cool down. We would often walk off our frustrations outside in the cold, sometimes even in the snow.

Ed and I were concerned that there were too many people working in the LM cabin. It was meant for two crewmen and was pretty tight when two of us were in it wearing pressurized spacesuits. Ed saw six people in the cabin a few times, which must have looked as comedic as seeing how many clowns can fit in a VW Beetle. The schedule pressure to overlap, manufacture, and test caused the overcrowded cabin. Some of the damages that needed to be fixed happened because the workers had limited elbow room and would inadvertently break the extremely delicate 26-gauge Kapton wiring. The use of this wire arose from the Super Weight Improvement Program that we called SWIP, which had the goal of trimming the craft's weight. Historically, all flying vehicles have faced weight challenges, but the LM was near the top of the list. Tom Kelly and Bill Voorhees probably looked at Ed and me as two nuisances, because we brought the overcrowding problem up at each status meeting. All we wanted was to come to consensus on the maximum number of people allowed in the cabin at one time. We eventually agreed to limit it to four.

Ed and I considered it a requirement to have either ourselves or a Grumman consulting pilot or cabin system operator tech in any powered vehicle, because someone had to be aboard to conduct an emergency power down of the vehicle if dictated. Several incidents occurred that reinforced the need for this coverage. A possible broken wire could cause an electrical short or a switch could inadvertently be bumped into and turned off or on.

My diary on April 25 says:

> I went out to the vehicle to look at panel installation and then the fun started. While looking at Panel 11 (Commander's side) I heard a gyro of some sort start to wind up over my head, the master alarm light lit up, and the flood lights came on. Then I turned to Panel 16 and had a real sinking feeling . . . almost all the circuit breakers were in.

I was referring to the installation of new instrument panels, during which time there should be no power on in the vehicle. I pulled the bus tie and DC bus circuit breakers, but the power was still on. I called the chief test conductor who was able to eventually power down. Dave Ballard found out that the portable temperature measurement unit, through the ASA circuit breaker, had a blocking diode missing.

On another occasion, while manning the vehicle cabin during testing, a tech called in that there was a loose wire on the starboard side of the space-craft causing sparks. I quickly dove out of the hatch and witnessed a dangling, loose wire hitting the side of LM-5, with sparks generated at each contact before bouncing away. This resulted in a stain being left, even after cleanup of the contacted skin area. I don't think that Neil and Buzz knew their LM had a beauty spot on the right cheek. The chief test conductor scurried to find a technician to locate the power source and turn it off.

Another nagging problem was the unkempt condition of the test stands. The ground support equipment, with associated cabling that was added for special external stimuli or data readings, was often left when that test was ended. As a result, much of this equipment created trip hazards, but also could be damaged by personnel stepping on the cables. Ed and I continually called this to the attention of the quality control employees on the floor and to Tom Kelly and Howie Wright. Walt Martens briefed me on a plan to install bridges between vehicle assembly/test stands 9, 10, and 11 for added space.

———

On May 10, 1967, Don Mallick at NASA Flight Research Center told me that Bruce Peterson was in an accident in the M2F2 lifting body on the Rogers Dry Lake at Edwards AFB. A week later, I visited Bruce at a hospital in Los Angeles. After rolling over several times on the lake bed, the vehicle was upside down when it finally came to a stop. The right side of Bruce's head and face was injured when his helmet came off. He was slated to be moved to the UCLA Medical Center for grafting a week from when I saw him. The doctors felt that his right eye had a fifty-fifty chance of fully recovering. As it turned out, an infection set in that required the removal of the eye. This ended Bruce's test pilot days, but he maintained his sense of humor, saying often, "I'm keeping an eye on you!"

Within a month, there was another accident. My office mate, Ed Givens, died in an automobile accident. On a rainy night in Houston, Ed had two Air Force reserve officers with him when his automobile slid off the road into a ditch. They survived. He was the senior Air Force member of our group, and I suspected that he would get an early mission assignment. The Original 19 had become the Original 18.

———

Driven by continual schedule slippage, Tom made a number of staff and orga-nizational changes. Lynn Radcliffe was a great addition to the management staff. He had worked at the test facility for the LM propulsion systems at White Sands, New Mexico. Lynn's capabilities provided some relief for Tom and Howard, who were getting worn down from dealing with the daily fire drills that could bring them back to work at all hours. NASA also had a management change with the appointment of Andy Hoboken as the head of the RASPO. Andy brought his experience from dealing with McDonnell test operations during the Mercury and Gemini programs. Lew Evans officially sanctioned the S/CAT organization and added spacecraft directors for each LM in process. Paul Butler was given responsibility for LM-2; Tommy Attridge, LM-3; Howard Miller, LM-4; and Al Beauregard, LM-5. They were senior and experienced having come from the Flight Test Department. All were very competent and, with their charter from higher management, could cut across organizational lines to break up any roadblocks in the flow. Paul carried his authority with clear logic and facts to get the cooperation needed, but Tommy bullied his way through roadblocks. We called him "Tommy Outrage," because it did not take him long to become highly charged. He was a former Grumman test pilot who was famous for having shot himself down while in a shallow dive when test firing the 20-millimeter cannon in the F-11F Tiger. Tommy simply overran his own bullets and subsequently survived a crash landing.

NASA had also assigned vehicle managers who reported to the LM program manager at the Houston MSC. They did not have day-to-day manufacturing/

testing responsibilities, but were brought on to record any problems and to assess how the schedule was progressing.

Unbeknownst to me, Ed Mitchell had volunteered to work on a project with the Grumman Research Department. It was the evaluation of the operator's ability to control a device, similar to a Segway or self-balance scooter, running on a metal track. The thought was this type of device may greatly increase astronaut mobility on the Moon. Ed casually mentioned to me that the research people had this new gadget and we were invited to try it out. I took the bait and followed Ed to a large room where the apparatus was set up for testing. Ed said that he would go first and with all his previous "secret" training, he easily ran around the oval track without a problem. Then I tried and tried and tried, falling off each time. Then out the corner of my eye, I noted the slight crack of a smile on Ed's face and knew that I had been had!

Even with all the turmoil going on, I was encouraged by a young tech in the LM cabin one night. He was repairing one of those fragile wires that had been broken. He said to me, "I don't know why you would want to go to the Moon, but I will do my best to make sure that the LM works well!" If the workforce at large had that attitude, I thought that we would surely deliver an exceptional LM-3 for Jim and Rusty to fly.

The first lunar module, LM-1, was delivered to KSC on June 22, 1967. Its first flight was on the S-IB Saturn rocket to Earth orbit. Typical of new vehicle deliveries, it arrived with problems. In particular, there were serious fluid and gaseous leaks coming from the propulsion systems. Similar issues with subsystem-component shortages had happened with the predecessor Mercury and Gemini programs.

George Skurla was the director of the Grumman Kennedy operation. He was a great guy who I got to know and respect, both during those challenging days at KSC, as well as when I worked for him when I joined Grumman after leaving NASA in 1979. George had many years at Grumman's aircraft operations, as an engineer and manager. But Rocco Petrone, who was the NASA director of vehicle launch operations, was not happy with the situation. I heard that he had called LM-1 a piece of junk. This did not sit well with Grumman, all the way to top management, because they were proud of delivering high quality aircraft, primarily to the Navy, for years. Rocco was a West Point graduate and

Ascent and descent stages, forming LM-1, are mated with the spacecraft LM adapter in the Manned Spacecraft Operations Building at KSC in November 1967. Courtesy of NASA

had a master's degree in engineering from MIT. George found out that it was not a pleasant experience to be dressed down by Rocco. He called for help from Bethpage operations, and Will Bischoff, subsystem engineer for structural and mechanical systems, was promptly dispatched to Florida to fix the problems.

LTA-8, the thermo-vacuum test article, was delivered to Houston on September 24, 1967. Jim Irwin and John Bull were slated to man the vehicle. They were to be suited in a series of twelve-hour tests in Building 32 at the space environment simulation laboratory. John Bull developed a medical problem and was replaced by Grumman test pilot Gerald Gibbons. Glennon Kingsley and USAF Maj. Joe Gagliano also joined Jim Irwin for the tests. All the test runs were completed on May 27, 29, and June 1, 1968. Ultimately, John Bull left the Astronaut Office on July 16 and had a long career with NASA at the Ames Research Center in California. The Original 19 was now the Original 17.

As LM-3 got closer to its move, I started working on the challenging chamber test ahead at KSC. In an Al Shepard memo on May 8, I was named to the KSC Altitude Chamber Board. I had some discussion with the LTA-8 team and asked them for the procedures that had been developed for their chamber test. I was particularly interested in the LM ingress of the crew. In August, I met with George Nelson, Grumman consulting pilot at KSC, about a ground-support-equipment system to build a set of umbilicals through a unique, LM upper hatch, not intended for flight, but for the crew to live off of until getting to altitude. To understand the complexities, some LM systems were not functional until the chamber was pumped down toward a vacuum. For example, a water sublimator cooled the equipment by forming a cake of ice in a vacuum that interfaced with multiple passes of glycol fluid. The timing of this had to be carefully controlled to allow the ice layer to develop. If too much equipment was powered up too soon, it produced too much heat, which broke through the ice layer.

Jim McDivitt requested the chamber test cover all of the LM's systems, both pressurized and unpressurized. Howie Sherman, of the crew systems group at Grumman that handled human interface with the spacecraft, first raised a concern about the crewmen having their helmet and gloves off during the chamber test. This concern was based on their remembrance of the LM-5 front window shattering during a pressure test in December 1967, at Grumman. The suggested compromise was to have one crewman with his helmet and gloves off when the chamber was taken down to 25,000 feet. The actual mission to land on the Moon could not be accomplished with this restriction. Howie, ultimately, lost the arguments, and the crew members, from time to time, had their helmets and gloves off during the lengthy tests that could last as long as seventeen hours. The procedural errors found as a result allowed for a second set of runs by Peter Conrad and Al Bean to go much smoother.

George Dowling and I were assigned to the manned drop tests in LTA-3. This process included drop tests to simulate the lunar landing at the estimated lunar touchdown landing weight of 14,673 pounds. A series of unmanned test runs were made before the manned test at 75 percent and 100 percent of the estimate, with the cabin pressurized. Because of the tight schedule, the manned test was reduced to one run, at 100 percent weight, to produce maximum tension at the forward interstage, and maximum lateral load at the aft interstage connection. George and I were onboard as "crash dummies" to experience the worst-case landing scenario that would certify the design of the crew-restraint system. The crew systems team worried about our safety and added extra padding around the spacesuit arms. George and I hooked up to umbilical hoses that supplied oxygen. We also latched the clips for the restraint cables to each side of our spacesuits. Cameras would capture our motion and blue, powdered chalk was on the interior aluminum walls, adjacent to our positions, to show where our bodies collided with the structure.

The vehicle was hoisted into the air above trap blocks that simulated lunar-crater landing conditions. The highest fall distance for touchdown above a trap block was twenty-three and one-half feet; the shortest fall was thirteen feet. The vehicle was released at the test conductor's command. The accelerometer mounted on the cabin camera recorded 6.3 Gs forward, 3 Gs to the side, and 6 Gs vertically. My diary for that day says:

> The initial drop load was about as I expected though the displace-
> ment was disappointing. I basically sunk straight down . . .
> maybe eight inches due to knee flexure. I didn't feel that I moved
> laterally to any great degree . . . either left or forward toward the
> window. The combination of locking up the reel assembly plus
> Velcro served to hold my feet flat on the floor at all times and
> firmly in place. The lack of lateral or forward displacement
> may be due to the much-abbreviated load time of approximately
> 25 milliseconds versus 150 milliseconds for the lunar case.

I do not know the sink rate at impact, but the landing we experienced was certainly much higher than any of the actual lunar landings.

A sell run on LM-3 of OCP 61018 was completed on May 11, and the vehicle was shipped to KSC in the Guppy aircraft on June 14, 1968. Rocco Petrone told George Skurla that he did not want any of those Bethpage people at KSC. He made it clear that he wanted George and his workforce to be his exclusive Grumman team at KSC. He had established a similar organizational relationship with Tom O'Malley, head of the North American Aviation group.

Testing got underway on LM-3, with problems that included bad flight subsystem hardware that needed to be replaced. This included the C-band beacons, the digital command assembly, the inertial measuring unit (IMU), the computer flight ropes (memory), the very high frequency (VHF) transceiver, and the hand controllers. The descent engine was to be removed to check for cracks in the nozzle, and a number of wires/connectors needed repair. The rendezvous radar had an electromagnetic-induction-interference issue that had gone unnoticed in Grumman New York's testing.

George Dowling and I suited up in the afternoon of August 1 for the manned altitude chamber test of the special ground support equipment umbilicals needed to support the upcoming LM chamber tests. I wrote in my diary:

> We entered the chamber at 4:30 p.m. and exited at 8:20 p.m. That suit really got heavy standing that long. A summary statement describing the gaseous oxygen umbilical system is that it supplies warm, breathable air at every delta pressure and flow rate. We found the optimum to be at 0.6 psi delta pressure and 12 to 14 cfm flow rate. Estimated air temperature at suit inlet was 70 degrees at best.

It was surreal that George and I, in our spacesuits, stood alone in that huge vacuum chamber cavity that would soon house the entire LM-3.

CHAPTER 8

MY TICKET TO THE MOON

I became the backup LM pilot for the Apollo 9 Mission (later to become Apollo 8) on August 8, 1968, which was planned to be the second Earth orbital mission with a LM. I would be joining Neil Armstrong and Buzz Aldrin. Frank Borman was the prime crew commander. Jim Lovell was the command module pilot and Bill Anders was the LM pilot. Ken Mattingly, Vance Brand, and Gerald Carr were the support crew. Joining the Apollo 9 Mission meant that I was the first of the competitive Original 19 to be given a flight assignment, but I felt Ed Mitchell won the competition because ultimately, he got to walk on the Moon first and I never did.

—

At a major program review meeting, chaired by Gen. Sam Phillips in the KSC headquarters conference room, all the top-level Apollo program managers were present, including George Low. The most important item on the agenda was a briefing by Charlie Mars on the LM-3 status. He shared his four-page LM-3 problem list, which precipitated a two-hour discussion that informed everyone in the room that the LM-3 would never make the planned launch schedule. George Low's additional seventeen issues really brought this home. He had already been considering an interim mission to the Moon, with just the CSM to fill the gap in the schedule. He discussed this possibility with Bob Gilruth and Chris Kraft, then sought the approval of George Mueller of NASA head-quarters to fly the alternate mission to the Moon, pending a successful flight test of the CSM on the upcoming Apollo 7 Mission.

As an official member of the Apollo 9 crew, on the first week in my new role, I flew to Los Angeles to participate in CSM 104 testing at the North American plant, along with Neil, Buzz, Vance, and Ken. This was my first time in the command module. What struck me were the stark differences in the CSM versus the LM. The CSM seemed spacious and had solid, metal inner walls, instead of all of the wiring bundles and plumbing exposed. The vehicle was not in flight configuration, with everything operating at the time, but sounded a lot quieter than the LM. I spent about eighteen hours in the CSM during tests of the electrical system.

I received a surprise at a pilot's meeting in Houston, when Deke announced that Apollo 9 was to become Apollo 8, which would fly to the Moon with only the CSM, pending the successful performance of the upcoming Apollo 7 Mission. Jim McDivitt's mission became Apollo 9 and retained the original objective of testing the first LM in Earth orbit. I was elated because in just two weeks, I had moved up to an earlier mission in the lineup. Then it hit me that there was no LM and I would have to learn how to operate a new vehicle. With the change of mission and impending launch only four months away, Frank Borman directed that our training schedule be reworked for six days a week, as opposed to the previous five.

My days were spent learning the systems in front of me, from the right couch position. These included the cryogenic tanks, glycol pumps, environmental control system radiators, cabin fans, inverters, fuel cells, entry and pyro batteries, and S-band/UHF communications. I studied the pertinent schematics and made use of the command module simulator (CMS), which was identical to a real spacecraft's interior geometry and layout of the instrument panel. The simulator could replicate normal system functions as well as impart a fairly extensive set of failures.

I spent a lot of time with Bill Anders, who I was backing up on the mission. I learned that he was also disappointed to find himself on this new mission without a LM. Many days, if we could get CMS exclusive time, we

The Apollo 8 prime and backup crew. Clockwise from back left: Frank Borman, Jim Lovell, Bill Anders, me, Buzz Aldrin, and Neil Armstrong. Courtesy of NASA

tested each other with a variety of failures to build our acumen for handling any problem. A nice feature of a simulator is if you screw up, you can ask for a reset and try again.

Bill realized that an important aspect of this mission was to photograph the exciting story of the first flight to the Moon. There were many meetings to devise and modify the photo plan, so that the most spectacular views would be captured. Ultimately, many incredible photographs were made, including Bill's iconic photo of the Earth rise over the Moon.

My first spacesuit fitting took place at the International Latex Corporation, thanks to Ken Mattingly, who found an A7L suit for me that only required arm and leg adjustments. Much of the workforce was composed of women who cut, stitched, and bonded rubber and cloth material to create the protective wear that was required. I didn't ask, but hoped the glue was super glue. The mission change caused many additional meetings to discuss the new flight plan; photo plan; mission rules; malfunction procedures; and guidance, navigation, and control. Bill Tindall, director of the Data Systems and Analysis Directorate, led a number of data-priority meetings. His detailed memos that we called Tindall-grams documented the meetings' key points and the associated resolutions. Normally, there were forty to sixty people in the conference room. Each one of them, based on their area of expertise, assessed the problems to come up with viable solutions. Fortunately, Bill was a master facilitator and kept order. He was adept at pulling every nugget of information from the brains of the attendees—even the shy ones—about subjects such as translunar navigation, thermal control, and manual takeover criteria during CSM entry into and out of lunar orbit. Bill's memos can be found on NASA's website. They should be reviewed by those planning new trips to the Moon or to Mars because they will face similar challenges.

I learned on a simulator, as with most of my previous flight training, because space flight was no different. The CMS was impressive, considering the mid-1960s technology. We laid in flight-configured couches, although we had the added comfort of padding that would normally be provided by a spacesuit. The instrument panel was an exact replica—all of the meters, warning lights, and sounds reflected each system as if we were in space. As we looked through the window, a visual of an Earth or Moon horizon and a star field was realistically displayed. For example, with the simulator set to run at any point in the future mission, the star field appeared exactly as it would during an actual mission. The simulator had approximately five hundred "credible" failures in the CMS and three hundred in the LM simulator that could be input by our training instructors. Credible failures were those that were survivable by crew and spacecraft. Explosions did not meet this definition. Frank Hughes was in charge of the CMS team and Charlie Floyd supervised the LMS team.

For the integrated flight simulation training with Mission Control, a group called Sim-Sup played a significant role. This group of talented, devious individuals, led by Jay Honeycutt, studied the suite of failures at their disposal and decided which ones were best to challenge the flight crew and Mission Control. Sim-Sup was overcome with pride when any of the flight team, alone or as a group, looked bad. At the completion of each run, the flight director led us through discussions on each failure and how the team handled it.

Being part of a crew was different for me. During my military career in single-seat fighters, I was a crew of one. Flying as a pilot or copilot during my hours in NASA multi-engine aircraft only involved teamwork during certain times. The training for Apollo 8 had us together every day, initially for six days a week, which, eventually grew to seven days a week in the last month of the schedule. These were often extremely long days. On my thirty-fifth birthday, I enjoyed a fourteen-hour stint. The day began with Jerry Carr reviewing and correcting the entry checklist document. I spent the afternoon in a spacesuit during a CMS session, covering launch through trans-lunar injection (TLI). I ended the day with a beautiful, starry, night race back to Houston with Jack Lousma. My diary on November 25 reflects how good I felt about working with Neil and Buzz as crew mates. "Neil is very competent technically and I've discovered he has a tremendous sense of humor. Sometimes his subtlety is just a little over my head. Buzz is our real thinker and, like Bill, goes into great detail over almost anything."

In meetings, Neil would not say much, but after analyzing the various points that others made about an issue, Neil would take the floor to share his concise, logical, best option. After working and observing Buzz, through both the Apollo 8 and 11 training, I came to realize that his out-of-the-box thinking was similar to that of the most advanced design people I worked with at Grumman years later. Buzz is a unique talent.

I could tell that Buzz was sometimes bored when we were in test mode in the spacecraft. This activity was like the OCP testing at Grumman: It was conducted through a formal procedure with rigid command and response. The action was slow-moving, and Buzz would pass the time working on the DSKY, or the Apollo Guidance computer's display keyboard. On one occasion, Buzz succeeded in freezing up the DSKY—something that was never supposed to happen. Draper Lab came up with a solution called Go Jam. This required two crewmen, with one depressing the PRO key on the DSKY while another one simultaneously depressed the Mark push button in the lower equipment bay. I don't know the technical magic behind this solution, but I assume it was a precursor to today's control-alt-delete.

Not all of our activities at NASA were related to flight. Our backup crew filled in at the launch pad so the prime crew could have lunch with Hubert Humphrey. The prime crew also left for Houston the following week to attend a Bob Hope benefit show at Jones Hall. As I found out later, when I was "up to bat" on the prime crew, there were many distractions that took time away from training, such as press conferences, media interviews, finalizing launch guest lists, and contacting family and friends for small items they may have wanted flown in the personal preference kit (PPK) on the mission. My favorite aunt asked me to fly her rosary beads, but because of the size and weight limitation of the PPK, I could only fly the crucifix.

——

We were the first crew to face the prospect of reentry from the Moon at 25,000 miles per hour. So we went through a few test runs in the centrifuge at JSC to assess our manual-control capabilities, to correct for translunar navigation errors. For example, at the lower or steeper edge of the entry corridor, the crew could experience up to 18 Gs. I found that, at the higher g runs, I started losing my peripheral vision, but I could still track the steering needle on an eight-ball instrument centered on the panel. Those high-G runs also gave me a rash on my back, caused by the small capillaries bursting from the severe pressing of my body downward into the couch.

Splashdown exercises in a boilerplate training capsule took place at the Gulf of Mexico and in a water tank at JSC. Training covered normal egress, with the capsule right side up, and when the capsule was upside down, which required an underwater exit through the upper hatch. Even with the brief time in the capsule in the relatively calm Gulf, I realized that it was not a very good, stable boat.

We went through emergency egress training on the launch pad at KSC. Many things could potentially go wrong—a catastrophic booster failure was one of them. We exercised several options of escape. One was a sliding wire basket that would take us away from the launch pad to scamper for shelter in a nearby bunker. A second escape route would be to take the elevator to the mobile launch platform level, where there was a round opening to a slide that led to the basement, to enter a rubber room. The room was circular and it was on springs. It had a door like a vault, which could withstand the rumbling and shaking of a potential booster failure. In the room there were water and oxygen supplies, along with a commode, to allow for survival until, theoretically, the rescue team dug down through the rubble to find survivors. During our first run, we realized the spacesuit did not allow one to slide smoothly down, so a modification was made to line the slide with Teflon. That made it fit for an amusement park experience.

Following the Apollo 1 tragedy and loss of the crew, many modifications were made to the command module capsule, including a new hatch design. The mechanism for quickly opening it was in reach of the command module pilot and the LM pilot. I could turn a switch-lever then grab a handle with a push button on the top. Depressing the button followed by three cranks of the lever, opened the hatch within ten seconds, with just a little training. I just hoped that I would never have to use it.

Following a routine flight physical, NASA physician Dr. Jorgenson informed me that I had a small shadow on one of my kidneys. In my mind, this was not just a potential issue regarding the Apollo 8 crew assignment—it could have had an even greater effect on my future flying career. I went through a number of X-rays and an intravenous series with dye, including thirty renal laminar X-rays that had me worried that I would glow in the dark. This was distracting, especially because we were now training seven days a week. On December 13, just eight days before launch, I found out that I had a tiny kidney stone, which was not an issue.

———

I became a member of Gunter Wendt's closeout crew, which was composed of six people. Our responsibilities were to ensure the astronauts were secure in their couch positions; to close and check for air leaks at the hatch; and to standby, at a halfway point to the Launch Control Center, to ready ourselves just in case we needed to rescue the crew. I went through fire training with the KSC fire department. One of the most exciting exercises involved donning a Scott air pack and crawling on my hands and knees through a burning structure. I presumed that the flames above and around the structure were fed by controlled natural gas. Gunter walked us through the locations of all of the equipment that we might need on the launch pad in case of an emergency.

The day before launch, I trained in the CMS with Neil until leaving for the eight-mile drive to the launch pad at 4:30 p.m. to perform the spacecraft switch list. This was the command-response ritual with the launch director, through 417 steps of the launch procedure. This process sets up the crew cabin for crew ingress the next morning. I departed at 9:00 p.m. to crew quarters to get a few hours of sleep. Leaving the launch pad, I admired the unbelievable sight of the enormous Saturn V standing there, bathed in spotlights. It was hard to believe that this beast would take Frank, Jim, and Bill to the Moon the next day.

When I awoke on my own at 2:40 a.m., I had butterflies. I heard some others moving around in crew quarters. I quickly showered and shaved. After my solo breakfast, I met with Bill, who was on cloud nine, then left to join Gunter at the LCC trailer rendezvous point, after going through two security checks on the road. Our close-out crew joined a small caravan headed to the glistening, smoking white monster that we could see ahead. I described it in my diary, "It was really a fantastic sight, rolling up on the pad with that towering giant covered with a thick layer of frost, while huffing and puffing out of numerous vents. And we were all alone . . . just the six of us!"

We took the elevator past the hissing cryogenic vents to the capsule level to the white room and walked down the same path the crew would soon follow. With the arrival of the crew, I climbed into the capsule and positioned myself in the lower equipment bay. As the crew climbed into the capsule, one by one, I handed the suit techs the restraint straps after they had transferred the hoses from the transfer oxygen canisters. Jim Lovell was the last one aboard, and I crawled underneath the couches to the hatch opening to crawl out. The hatch was closed and the North American Aviation member of the team conducted a seal check. I was amazed because the crew that I had worked with nearly every day over the past few months was about to be launched to the Moon. I had never witnessed a Saturn V launch. The ignition flame was so incredible that I wondered what would be left of the launch pad. As the rocket moved slowly upward, the fury of the sound waves hit and the fire-retardant clothing that I was wearing began flapping from the sound-pressure waves that I felt deep in my chest. I considered what it felt like to be in the capsule and remembered that I would soon experience it firsthand.

—

I was not aware of the crew's plan for their Christmas Eve reading, but thought it a perfect historical footnote for the mission. On December 29, Mary and I went out to Ellington with Jack and Gratia Lousma to greet the crew on their return at 2:00 a.m., along with several thousand others. I was astounded, not just from the crowd at Ellington, but from following the media and worldwide news of the mission. Somehow, being so wound up in the nuts and bolts training, I had never considered the public's reaction. To me, this was merely another great adventure, and from my rudimentary understanding of astronomy, going to the Moon was not really going all that far.

I took a week's vacation to recharge my batteries, just spending time at home with my family. Bill Anders let me know that our new assignment was with Jim Lovell as the backup crew for Apollo 11.

The Apollo 11 training ahead offered a few new things to prepare for, and now I was back with a familiar friend—the LM. It was the LM-5 that I spent quite a few hours testing at Grumman in 1967. Traveling to the Moon, getting into its lunar orbit, and reaching home through a Mach 36 entry are all of the mission phases that had been proven valid with Apollo 8. However, Apollo 11 covered new training for LM operations at the Moon, including landing on the Moon, subsequent extravehicular activity (EVA), and walking on the lunar surface. Then there was the first launch of the LM ascent stage, with a rendezvous sequence to return for docking with the mother ship.

Since the training plan had not yet been published, I spent the first month as a backup crew of one. My crewmates Jim Lovell and Bill Anders were on a whirlwind public relations tour, including ticker tape parades in New York and Chicago. They also attended President Nixon's inauguration.

I looked forward to flying the lunar landing training vehicle because of the potential that I would enjoy it, but management decided that only commanders on designated landing missions would fly the vehicle. My turn came when I served as Apollo 16's backup commander.

Bill left the training cycle for Apollo 11, despite the fact that he wanted to make a landing on the Moon. I think he figured that he would never pass through the mission cycles to have that opportunity. I also think Bill was smarter than I about the political world and its conceivable effect on NASA's budget. No one really knew if the space program would always be funded at the same level. Bill became the executive secretary of the National Aeronautics and Space Council. Ken Mattingly, my old roommate, took Bill's place as a command module pilot.

The LM simulator was new ground. Any spare time that I had I trained in it, asking the simulator instructors to hit me with failures. Even though I had witnessed actual problems during my testing of LMs at Grumman, this was the first opportunity I had to look thoroughly at the entire catalogue of things that could possibly go wrong. In the simulator, the view for landing, as one looked out of the window, was a large plaster of Paris mold of the landing area on the Moon. Since the Moon's surface was covered with many boulders, it was necessary for the commander to manually land the craft as a safety measure, despite the onboard computers that programmed the landing site. The molds were designed from images collected on the five lunar-orbiter missions. Another novel experience was flying the KC-135, also known as the "Vomit Comet," which was used in evaluating the setup of lunar surface experiments. This was my first exposure to the one-sixth g of the Moon's environment.

We had training in the proper procedure for gathering Moon rocks, which took place at the Quitman Mountains in Texas. This activity also added to our knowledge of various rock types. There was an abundance of volcanic rock specimens in this area. The objective of the sessions was to reinforce the collection protocol, covering photography methods and labeling. Each sample was to be photographed twice, with a sidestep in between the snapshots, to create a 3D effect. The sample had to be carefully bagged and sealed.

As usual, with every space mission up to that point, the training schedule got more intense as we approached the launch date. On all four of my mission assignments, the training in the last month went on seven days a week, from eight in the morning until ten o'clock at night. It was hard on Mary being the only parent at home, but we had to be focused. As a matter of fact, I had been asked on many occasions if I was working so intensely to make sure that the US beat the Russians. Well, I had no idea of what was going on with the Russian space program. In fact, I hardly read newspapers or listened to the news, so I had little knowledge of what was going on in the world. I felt like I was in an alternate universe, spending time in simulators and meetings and discussions that no one outside of our environment could understand or appreciate. Case in point: One evening, Neil, Buzz, Mike, and I met at Ramon's Steak Restaurant in Cocoa Beach for dinner. Neil and Buzz got into a loud disagreement over the REFSMMAT, or reference to stable member matrix. (I am sure that cleared things up for you.) They were discussing what the eight-ball attitude instrument would look like if the landing was made at the computer-designated landing spot. This was pertinent because Tommy Gold, a respected astrophysicist at Cornell University, suggested that the Moon's powdery, dusty surface layer would be stirred up by the descent engine just before shutdown. The dust storm created would require an instrument landing. Neil and Buzz's gobble-dygook drew some odd looks from the other patrons. I suspected that those onlookers who lived in the real, everyday world had no idea of what those two guys were arguing about, but Mike, as he frequently did, interceded to calm Neil and Buzz down.

I volunteered again to be a member of the close-out crew for Guenter Wendt, or as we fondly called him, "der Fuhrer of der Launch Pad." This was the mission that hopefully would fulfill the Apollo program's hopes of landing on the Moon. On the day of the launch, I think that many of us remembered President John F. Kennedy's words in front of Congress on May 25, 1961: "I believe that this nation should commit itself to achieving the goal, before this decade is out, of landing a man on the Moon and returning him safely to the Earth." But, perhaps, his speech at Rice University, on September 12, 1962, had more meaning for those of us working between twelve to sixteen hours a day to win the Space Race:

> We choose to go to the Moon in this decade and do the other things, not because they are easy, but because they are hard, because that goal will serve to organize and measure the best of our energies and skills, because that challenge is one that we are willing to accept, one we are unwilling to postpone, and one we intend to win and the others, too.

———

Once again, I witnessed Saturn V emitting white smoke and hissing through its vents. This rocket was alive and ready to go, and, as I did on the previous Saturn V launch, I took my post in the lower equipment bay and assisted the suit techs in getting the crew—Neil, Mike, and Buzz—switched over to the spacecraft suit hoses from their transit oxygen canisters and getting their restraint harnesses in place. When my duties were done, I gave the crew a thumbs up and crawled under the couches to exit out of the hatch. Just before entering the spacecraft, Mike had presented Guenter with a present of a real fish as a joke, holding it with his spacesuit glove. I could see the headlines when fish scales were found in *Columbia* during the postflight inspection.

Later that day, I heard that the TLI had been successful and the crew was coasting on their way to the Moon. All was going well. As a backup, Jim Lovell and I were in Mission Control during the flight just in case anything went wrong. The tension and excitement grew in the room during the actual landing, but when a number of computer alarms sounded, I worried about an abort. I am sure that most of the other people sitting in Mission Control on that day felt the same apprehension. Charlie Duke was the capsule communicator—CapCom for short. He spoke to Buzz and Neil in the LM as they landed on the Moon:

Duke: Eagle, Houston, you're go for landing, over.
Aldrin: Roger, understand. Go for landing. 3,000 feet. Program alarm.
Duke: Copy.

Aldrin: 1201.

Armstrong: 1201.

Duke: Roger. 1201 alarm. We're go. Same type. We're go.

Aldrin: 2,000 feet. 2,000 feet. Into the AGS, 47 degrees.

Duke: Roger.

Aldrin: 47 degrees.

Duke: Eagle, looking great. You're go. Roger, 1202. We copy it.

Then, two-and-a-half minutes later:

Duke: 30 seconds.

Armstrong: Forward drift?

Aldrin: Yes. Okay. Contact light. Okay, engine stop. ACA out of detent.

Armstrong: Out of detent.

Aldrin: Mode control—both auto. Descent engine command override off. Engine arm off. 413 is in.

Duke: We copy you down, Eagle.

Armstrong: Houston, Tranquility Base here. The Eagle has landed.

Duke: Roger, Tranquility, we copy you on the ground. You got a bunch of guys about to turn blue. We're breathing again. Thanks a lot.

Charlie Duke's last statement, with his South Carolina accent, perfectly described the emotions in the control room. Neil had not reported any vehicle control problems with the alarms, and I was hoping that an abort would not be called prematurely. *Never panic early!*

I left Mission Control to watch the EVA with Buzz's family at his home in Nassau Bay. I had a copy of the checklist the crew would follow to share with his wife, Joan, and their two sons and daughter. I had to hurry because those men on the Moon decided to skip the planned rest period and get on with the Moon walk. I knew their adrenaline was flowing during that landing. As I watched the fuzzy images of Neil and Buzz shuffling around

Charlie Duke, Jim Lovell, and me at the CapCom console during the Apollo 11 landing. Courtesy of NASA

on the lunar surface, I wished that a better TV camera had been developed to capture that historic action.

When they returned, they were quarantined in a sealed glass area that was part of their home-away-from-home for almost three weeks. I, along with many of my Astronaut Office colleagues, met with them in quarantine to hear their firsthand debriefing of the flight in the lunar lab. The quarantine was required because they may have brought new bacteria or a virus from the Moon that could have threatened the population on Earth. They were the first crew to go through the quarantine protocol that started with donning the biological isolation garments, while still in the *Columbia* spacecraft after splashdown. When they boarded the aircraft carrier, they entered the mobile quarantine facility for physical examinations by a medical officer. President Nixon visited them there—he had also wanted to have breakfast with them before their launch at KSC, but NASA physician Dr. Berry would not allow it. The crew rode in their special "vacation" trailer, which had accommodations that could be found in an RV, as well as an examining room to perform postflight medical evaluations.

The worldwide media coverage was amazing. I never imagined the extent of the interest. New York and Chicago threw ticker tape parades. The first-ever State Dinner outside of Washington, DC, was held at the Century Plaza Hotel in Beverly Hills. I was at a table with Ken Mattingly. He asked Rosalind Russell, who was seated with us, what she did for a living. At least I knew she was a movie star. Since the dinner was so close to Hollywood, many celebrities from that glamorous business attended. There seemed to be no end of US requests for the crew to appear and be lauded by throngs of people. After visiting the United Nations, the crew was dispatched to tour twenty-two countries. I hoped that all this notoriety and fanfare would die out by the time I landed on the Moon.

———

CHAPTER 9

ODYSSEY— A PERFECT NAME

Jim, Ken, and I were assigned as prime crew for Apollo 13. So, it was back to the training routine for my third cycle, but this time I would be rewarded with flying to the Moon. The backup crew was John Young, Jack Swigert, and Charlie Duke.

After the Apollo 11 Presidential State Dinner in Beverly Hills, I stayed over to train with Ken at the Griffith Planetarium. This was a refresher on finding the constellations stored in the spacecraft computer. Planetarium personnel would spin the star field on the dome and test us to see if we could figure out where we were from the constellations that were visible.

Our training coordinator, Lloyd Reeder, scheduled me to try out a new tool for use on the lunar surface—the lunar drill. The plan was to collect a core sample from the Moon with this tool, which could drill down ten feet. I found drilling was a difficult task in the pressurized spacesuit and the clumsy gloves.

We climbed into *Odyssey* for the KSC altitude chamber test. Jim and I enjoyed a great barbecue dinner with a side of baked beans at Fat Boy's restaurant in Cocoa Beach the night before the CSM chamber test. The food was delicious, but didn't settle well—I think Ken wished he was somewhere else during the test because it takes several passes through the suit loop and filter to clear odor from the air.

Sim-Sup worked hard to keep the team, crew, and Mission Control challenged through the myriad of integrated simulations over the months. One would think that after training for two prior missions, I had seen everything, but when I tried to outthink Sim-Sup to guess what failures were coming, I was often wrong. Every simulation was a new experience. However, as the launch date approached, we ran segments of flight without any failures to get a feel for the normal mission timeline, without the interruption of things going wrong.

In events with the general public, I am often asked how we seemed so controlled in handling the inflight failure on Apollo 13. They were not aware of our training, where dealing with failures was business as usual. Ron Howard, doing his homework before filming the movie *Apollo 13*, said that he listened to all the air-to-ground transmissions provided by NASA and it never seemed to him that we had a problem.

I talked to Jim about expanding the field geology training based on a conversation that I had with astronaut Jack Schmitt. Because the launch was rescheduled, we had more time for geology training. Jim agreed to join a meeting that Jack Schmitt had set up in Cocoa Beach with Lee Silver, a Caltech professor. Lee is a gregarious fellow who is easy to like the first time one meets him. He offered Jim an opportunity to lead us in a trial exercise. We joined Lee Silver and his associate, Tom Anderson, in the Orocopia Mountains Wilderness area in California for the additional geology field training. In the summer, this is one of the hottest parts of the country. However, the weather was fairly mild for our three-day visit, so the plan was to sleep on folding cots in the open and do our own cooking. Jim and I, as well as our backup crew of John Young and Charlie Duke, took part. Lee had us doing two exercises a day using Polaroid cameras to record our sampling and observations. We continued to talk geology into the night around the campfire. One evening, Charlie Duke volunteered to cook some of his good South Carolina fried chicken, but there was a secondary effect, because the frying pan was not fully rinsed of the detergent that had been used to wash it. By the next day, several of us had a case of "detergentitis," so Tom Anderson had to make an emergency run to civilization to get a supply of toilet tissue. This experience earned Charlie the nickname "Typhoid Mary."

Jim and I left the Orocopias for Los Angeles, with plans to attend the Society of Experimental Test Pilots black-tie awards banquet at the prestigious Beverly Hills Hilton Hotel. We had not showered for several days, were in our dirty field clothing, and had a case of dark whiskers. Looking like bums off the street, we strode up to the desk to check in. Some of Jim's humor came out when he said to the clerk, "This is a stickup!" Her eyes widened and she seemed to be in shock. I quickly spoiled the fun by showing my ID and explaining our appearance, because I did not want us to have to explain the holdup threat at the police station, which would have delayed us considerably.

Five more geology field trips were included in our training schedule. The last one took place at the Black Canyon Crater Field near Verde Valley, Arizona. This was three-and-a-half weeks before our launch. These exercises evolved into mission-oriented simulations. A set of traverses, or planned routes, were planned for each training site, with the geologists who were going to be in a room across the hall from Mission Control during the actual EVAs on the Moon. The mission geologists joined us on the trip and operated from a tent near our training area. To prevent an unfair advantage, if any of the geologists had ever studied the particular area in their past, they were excused from participating.

Lee Silver and an available geologist who knew the area followed along to critique us and the other geologists on what we, collectively, had failed to observe and sample. Jim felt that our field training qualified our mission to be the first that officially prioritized science. He worked on the Apollo 13 patch design with Lumen Winters, a New York artist. The motto sewn into it boldly stated, *Ex Luna, Scientia* or "From the Moon, Knowledge."

We had many exercises at KSC on the facsimile of the lunar surface. The training was done wearing our spacesuits—these days were grueling. We started our day in the cabin of a full-scale LM mockup, going through the procedure of suiting up, donning the portable life support system backpack, and getting pressurized. Some days we were suited up for six hours. There would be a halt to depressurize and remove our helmet and gloves to then climb down the ladder, because on Earth it is unsafe to climb a ladder while pressurized. After donning helmet and gloves to pressurize our spacesuits once again, we would go through all the procedures spelled out on our cuff checklists, for activity around the deployed modularized equipment stowage assembly. On EVA-1, I had a task to extract a mockup of the SNAP 27 isotopic power system for the experiment package. The handle to remove the device had a shield to protect your hands from the heat of radioactivity when handling it on the Moon. We would then, once again, depressurize to hop onto the tailgate of a pickup truck for a short ride out to the lunar field. Suit techs accompanied us while we followed the steps on our cuff checklist. They carried an ice chest with them that was configured to supply cooling water to tubes immersed in our undergarments, because the heavy spacesuits, when worn on Earth, are so hot. Despite the cooling water, we were soaking wet from perspiration that had worked its way down into our lunar boots.

A new training experience was scheduled in the JSC centrifuge facility. Some of the engineers figured out how to suspend a trainee in a pressurized spacesuit to simulate one-sixth G. We practiced walking or hopping around a circular track. The terrain was covered with sandy soil to replicate the Moon's surface. This training was great for practicing one's mobility, but the drawback was the painful pressure points from being suspended in the spacesuit.

We spent a day at the JSC food lab to sample the freeze-dried, powdered food options that we would take on the mission. That included food such as spaghetti, beef and vegetables, hot dogs, and snacks like bite-sized bread cubes, cookie cubes, and peanuts. The grub was not as good as Mom's home cooking, but it was better than C-rations.

Another exercise was scheduled in the JSC water tank to go through the contingency transfer from the LM to the command module. This would be required if the tunnel could not be used for mobility between the two docked spacecraft. I hoped that the simulated failures to create this situation would never occur, because it would require the complex, risky transfer of ourselves and the precious lunar rock boxes across the exterior of the two spacecraft.

Through all of our training, I was in a Space Race bubble. I was so locked into our training and mission that little news of the outside world filtered in, but it was a volatile time. That April of 1970, the Vietnam War raged on and a Green Beret outpost was attacked during a siege in the Central Highlands region, northeast of Saigon. Florida's governor defied a federal judge's orders for school integration in Tampa. The official US Census reported on April 1 that the population of the United States was 203,392,031. "Let It Be" by the Beatles was number one on the charts, but, on April 10, Paul McCartney left the band. On April 7, John Wayne won his first and only Academy Award for Best Actor in the film, *True Grit*. President Nixon signed a bill into law banning cigarette ads on television and radio.

We had our last press conference before launch on March 14. Every time I saw our Saturn V bathed in floodlights at night, I became more and more excited. People sent me letters to express their concerns and ask me if I worried that I was on mission 13, because that was an unlucky number. I was never superstitious, so I filed the letters in the wastebasket. But a few days before the launch, Charlie Duke reported that he had been exposed to measles at a birthday party with his young son. Since both prime and backup crews were always together, we were all exposed. As a result, all of us had blood tests on several mornings. The week that should have been spent looking forward to our flight to the Moon turned out to be one of worry.

Three days before liftoff, NASA physicians decided that Ken Mattingly should not fly. He had never had the measles, unlike Jim and me, and was therefore deemed to be at a higher risk of infection. So, for the first time in the Mercury, Gemini, and Apollo programs, there was a crew change. Jack Swigert replaced Ken, who was tremendously disappointed. People have asked me whether this last-minute crew change had me worried about mission success, but it didn't because I had served as a backup twice. I was completely confident in everyone's readiness, because we all went through the same training.

The crew change was painful and unfair to Ken and Jack. Ken's family and friends were all prepared to attend the launch—airline tickets were purchased and motel rooms reserved. Jack, on the other hand, did not have the opportunity to coordinate getting all of his people to attend. The PPK situation suffered, too. It took me several months before the flight to canvass family and friends for the small items they wanted to be flown. Ken's PPK was packed on the *Odyssey*, and there was no time for Jack to prepare his. There was a mix of emotion ranging from sadness, elation, and contentment to frustration.

The new crew of Apollo 13: Jim Lovell, Jack Swigert, and me. There was no time to don our spacesuits after the last-minute crew change.
Courtesy of NASA

Normally, in the few days before launch, the crew would relax and review the flight plan. There was a nice beach house for quiet time. There was no phone there to assure that we would not be interrupted. The house was a remainder from a small town that was closed to build the KSC. Because of the crew change, we didn't get to experience the calm.

After the crew change, Jim thought it prudent for us to work in the simulators to go through each of the dynamic phases of flight, including launch, lunar orbit insertion, trans-Earth injection, rendezvous, and entry.

We were in the simulators until 8:00 p.m. the night before launch. One simulated test failure on the LM required Jack to complete the rendezvous from *Odyssey*.

On launch day, I woke up feeling excited and remorseful because Ken would not be with us. Deke Slayton was the only one with us at breakfast. We had the traditional launch-day menu of steak, eggs, toast, coffee, and orange juice. Conversation was uncharacteristically subdued. From there, we went through the routine preparations that occurred before suiting up. From spacesuit dressing and the attachment of medical leads onto my chest to allow Mission Control to read my vital signs, to being strapped into the spacecraft couch, the thought that kept ringing through my mind was "I sure hope I don't screw up."

I could not hear anything in my spacesuit as I took the elevator up to the walk across the swing arm to the clean room. All of this was familiar to me from my two previous stints as a member of the closeout crew for Apollo 8 and 11.

Securely settled in the right couch position, I had nothing to do except wait. When I felt a slight motion of the stack from the gimbal test of the four outer F1 engines, I knew we were getting closer to liftoff, and the next stage would be the retraction of the hold-down arms. As we took off, the pressure increased, along with an increasing g level. We were pushed by 7.5 million pounds of thrust from our five rocket engines. The most unusual sensation I felt was a pronounced herky-jerky, left-to-right motion, caused by the gimbaling of the four outer F-1 engines, to keep the rocket following the programmed path. The motion was probably exaggerated because we sat in the very top of the 363-foot rocket.

A newspaper article captured the words of my wife, Mary, who was at home watching the launch on television. She was pregnant with our son, Thomas. "It's trite to say it was a marvelous launch," Mary told the reporters, "but it was a marvelous launch." She also told them it was a beautiful sight.

After two minutes, the center F-1 engine was automatically shut down as planned to keep the maximum G level from exceeding 4.5. Fighters that I had flown were capable of 7.33 Gs, so the acceleration that I felt riding the rocket was not that impressive. Although I had been briefed by previous crews, the deceleration after the shutdown of the four engines was still surprising. I was glad that I had been strapped in so tightly to prevent me from hitting the instrument panel. After we were separated from the first stage, five J-2 engines with 1,150,000 pounds of thrust cranked up to settle us back in our couches.

At three minutes and twenty seconds, after Jim jettisoned the launch escape tower, the view I saw far exceeded anything I had seen from an airplane. As we continued to climb, I was amazed by the incredible sight of the Earth as I viewed the horizon. It seemed unreal. I felt an increasing, then a decreasing g level, with a brief high-frequency vibration, followed by a slight drop in acceleration. This was another one of those *never panic early* moments. As Jim reported a center-engine shutdown, I noticed that the number five engine light had gone out on the instrument panel in front of him. The master alarm light came on and we heard the alarm squealing in our headsets, but everything felt normal otherwise. I was primarily worried that this early engine trouble would lead to an abort, with us ending up on an alternate Earth orbit mission.

Jim asked Mission Control for a reading on the engine and received a response that we were "Go," with no explanation for the shutdown. Jim pushed the master alarm button to stop the squealing in our ears. The subsequent S-II and S-IVB stage performance was normal, with an orbital insertion at twelve minutes and twenty-nine seconds, with our altitude right at one hundred miles.

After doffing my helmet and gloves, I tackled my assigned tasks with vigor. At this point all I wanted to do was look out of the window, but I was assigned to unstow equipment to take pictures. The cameras, both a Hasselblad and a TV camera, as well as film packets, cabling, and window brackets, were in storage containers under the couches. I had been cautioned to move slowly for a while to get acclimated to zero G. Unfortunately, I was not only moving too fast, but I also had to rotate and turn to get to the storage lockers below the couches. This caused me to spit up a couple of ounces of fluid into a bag that I had in my suit pocket. After that, I strapped myself loosely onto my couch until the malaise went away. I learned my lesson and from then on, I moved slowly.

As we approached Baja California and then when we were over Africa, I began shooting. The view of the Earth was incredible. I saw lightning in storm clouds and a few lights in Africa that was otherwise almost completely dark. The splash of white on California was the snowcapped range of the Sierra Mountains. After TLI, and with the full planet in view, I saw land masses and full continents. In vivid colors and hues, as rich as Technicolor, the blue of the oceans and streaks of white clouds were astounding. It was surreal to shoot stills of the shrinking Earth for several hours. I have found it hard to describe the deep emotion I felt viewing these out-of-this-world scenes.

—

The third stage was ignited again for the TLI to accelerate and settle us in our couch. After the single J-2 engine burned for six minutes, we gained escape velocity. Shortly after that Jim and Jack changed positions to allow Jack to control the docking with *Aquarius* that was still nestled in the S-IVB. I made sure that I had both cameras ready to chronicle the final closure. Jack continually called his distance away from *Odyssey*. CSM vehicles were not equipped with radar, so this docking, as on all missions, was done manually. *Odyssey* approached the LM *Aquarius* very slowly and, at impact, I felt it, but it was not jarring. Two of the latches that held the two spacecraft together did not initially lock, and we had to cycle through again to get all to latch. Jack backed our now combined vehicles away to clear the S-IVB. Mission Control Center (MCC) directed a booster attitude change and a small S-IVB Auxiliary Propulsion System maneuver of about ten feet per second for separation after Jack reported being visually clear. With greater separation, MCC vented the S-IVB to alter its path to assure impact with the Moon. This was the first mission that this was done and provided data to be collected by the two seismometer instruments left during the Apollo 11 and 12 missions.

With some breathing time, we carefully removed and stowed our spacesuits. The cabin seemed a bit roomier after we got all dressed up in our two-piece beta cloth garments.

Our first use of our *Odyssey* 20,500 pound thrust service propulsion system (SPS) engine was for MCC-2 at 30:40:30 hours into the mission. That thrust level was slightly more than the Pratt & Whitney J-75 at military thrust level in the F-106 Delta Dart that I had flown. The best characteristic of the SPS rocket engine was its reliability. In space travel there is no margin for "iffy" equipment without backup, because if the SPS engine failed after going into lunar orbit without the LM available, the crew would be stranded. That was a concern every mission faced. The 3.7 second firing provided less than a one-G acceleration and served to put us on a different trajectory. We were now on a path that would not provide us a turn around the Moon and back to Earth for an entry. As it turned out, this was our only use of the SPS engine on the Apollo 13 Mission.

Another one of those things overlooked in the hectic week leading up to the launch was Jack livening things up when his request for an extension on his IRS tax filing made the news. After the chuckles in Mission Control, Joe Kerwin called back to let Jack know that the IRS had granted him a sixty-day extension to file because he was "out of the country."

For the next two days, our activity mostly involved housekeeping. Jim and Jack conducted P-52 program star alignments, P23 cislunar navigation, and setting up the slow rotation with passive thermal control. The latter was to set up a rotation of the combined spacecraft to assure an equal exposure to the sun on all sides. I conducted some of the more mundane tasks of changing out lithium hydroxide canisters, purging the fuel cells, and doing water dumps because the fuel cells produced water as a by-product and would fill the tanks.

The schedule included sleep and eat periods. Preparing the dried, powdered meals took a lot of time to completely mix the hot water with the powder. By the time it was ready to eat, the food was no longer hot, so I used our gray duct tape to attach the food bag to a flood light to reheat it.

We conducted a scientific experiment that recorded our observations of how cosmic rays impacted our eyes in a totally dark cabin. With some concentration, I noted occasional small white flashes of light in my eyes, but nothing else was unusual.

During my free time, I enjoyed the intriguing view as *Odyssey* rotated. The Earth got smaller and smaller, as the Moon grew larger. CapCom Joe Kerwin called us at one point to tell us that they were bored to tears.

———

CHAPTER 10

A SUDDEN DETOUR

Previous mission TV shows only presented the interior of the LM. So, Jim and I decided that we would have a show-and-tell of the equipment not covered before. We started in the CSM cockpit, then floated down into our LM *Aquarius*. From his perch on the ascent rocket engine cover, Jim described the interior of the LM. I explained our backpack storage and the oxygen purge system, also known as OPS. Before taping, we set up our sleeping hammocks to be used on the lunar surface, but when the film was rolling I had difficulty getting to bed in zero G. After that, I discussed the water supply bag that had been added to our spacesuits. It had a small hose and mouthpiece that allowed us to get a drink while on the lunar surface, because Pete and Al had complained about getting thirsty on their Apollo 12 EVA.

Our TV show dragged on and on because I did not have enough prep time to get some of the items out of their storage locations. The only filming that we spent a lot of time preparing for was the one slated for lunar orbit. Of our crew, Ken Mattingly spent the most time with Farouk El-Baz, our lunar scientist, to write the script describing various lunar surface targets and the history of their names.

As I closed out the LM, I heard a loud bang echoing through the metal hull. The attitude jets were firing, and I felt the vehicle move. This was at

The live television broadcast on the way to the Moon wasn't carried by any network. Mission Control seemed relaxed, up until the explosion took out one side of our service module. Courtesy of NASA

55:54:53 mission time. As I floated back through the tunnel to get to my couch position in *Odyssey,* Jack calmly stated, "Houston, we've had a problem here." Houston did not respond to Jack's call, so Jim repeated it. The CSM and LM were torqued by the force of the explosion. I heard the metal structure bending in the tunnel that connected them. It sounded like an empty pop can being squeezed. When I got back to the CSM *Odyssey,* I noted a number of caution and warning lights. Most of them were connected to the electrical system. There was a Main B Undervolt FC 1, AC Bus 2, AC Bus 2 Overload, Restart, and a red Master Alarm light.

Jack fired the attitude jets to avoid an attitude excursion into IMU gimbal lock. The situation was clearly one of *never panic early.* I was confused. There was no single failure that could cause these problems in the various, standalone systems. In our simple spacecraft, the systems were mainly independent. And, through all of our training, the Sim-Sup group had never planned more than one failure at a time when devising scenarios to test us. I think the philosophy was that God would not be so unkind as to give us more than one failure in one system simultaneously.

In the case of Apollo 13, Murphy's Law prevailed: "What can go wrong will go wrong." The needles for oxygen tank 2, for both pressure and O_2 quantity, read zero. I knew it was highly unlikely for two different sensors to have failed at the same time. I also knew that the loss of one oxygen tank would call for an abort, and I was sick to my stomach, because I realized we had lost the landing and would not even be cleared to go into lunar orbit. I felt that all of my training preparing to fly on Apollo 8, 11, and 13 was a waste.

I did not think that the situation was life threatening, because it appeared we had a fully functional oxygen tank 1. I figured Mission Control would have us return home. Jack and then Jim attempted to install the command module hatch, worrying that we might lose cabin pressure. From their knowledge and confidence in the CSM, they thought that the problem was with the LM. On the other hand, from my intimate knowledge of the LM, I assumed we had a problem with the CSM.

When Jim, looking out of the hatch window, reported something venting from *Odyssey*, it became apparent to Mission Control that this was a real problem and not merely one of malfunctioning lights or alarms. Tackling the AC Bus 2 Overload, I started to switch some system components over to AC Bus 1. This resulted in triggering a Main A Undervolt light. The fuel cell readings were even more confusing. Fuel cell 1 had no flow and there were similar readings on fuel cell 3, and it appeared that all of the fuel cell reactant valves were open. The question that kept running through my mind was how could any single failure cause the simultaneous loss of two fuel cells? I had not realized that our panel indicators only changed position if both the oxygen and hydrogen valves changed position. Mission Control had telemetry that showed the status of both valves, but did not inform us that the oxygen reactant valves had moved to the closed position when we lost fuel cells 1 and 3. Of course we had no way of knowing that the springs attached to the valves were overcome by the g shock of the explosion. The g levels recorded were 1.17 in the x axis and 0.15 in y and z. Eventually, the crew onboard and Mission Control realized that pressure was slowly decreasing in oxygen tank 1, so we worked with

Mission Control through the next hour, troubleshooting to try to stop the leak in our remaining tank. When Jack Lousma called to tell us to shut off fuel cell reactant valves, I asked him to repeat the request, because I knew that action was irreversible—the fuel cell could not be restarted. He also asked us to isolate the O_2 Repress, making it obvious that our mother ship *Odyssey* would be lost, so Jim asked me to head to *Aquarius* to start a power up.

More than twenty-five years later, one afternoon, after retiring from Northrop Grumman, I decided to listen to the various voice loops in Mission Control during the two hours and forty-six minutes, from the oxygen tank explosion to Jack's decision to shut down the *Odyssey*. From my training through three missions, I knew many of those involved in the various disciplines by their voices. I detected a change in tone from some of the men when they had run out of ideas.

Flight director Glynn Lunney and his black team had replaced mission director Gene Kranz and his white team shortly after the explosion. Lunney called the team to attention with the call to get *Odyssey* powered down because we were starting to eat into the three small forty-four-amp-hour entry batteries. This was unprecedented, because there was no procedure to shut down the CSM *Odyssey* since no one ever thought it would be necessary. It was fascinating to hear the men of various disciplines discuss the "unknown effect" of shutting off system elements. It was clear that they had not given up on us getting back to Earth.

Jim joined me in *Aquarius* with Jack Lousma still on duty at the CapCom console. I called back and forth with Jack to inform him that I was bypassing certain steps for expediency, and because some systems were not needed in our current situation—like the landing or rendezvous radars. A very important step in the procedure was to transfer the Noun 20 gimbal angles displayed on the keyboard from *Odyssey* to provide our *Aquarius* IMU with that precise attitude reference. This was critical to performing any subsequent engine burns to alter our trajectory. Jim wrote the angles down that Jack passed and made a manual correction for the small angle misalignment noted in the tunnel from our docking. Because of the importance of the data being correct, Jim asked Mission Control to validate his numbers. Before I made each entry into the keyboard,

Floating through the tunnel into the LM where we lived for more than three days until releasing it just before reentry. Courtesy of NASA

I stopped to allow Jim to confirm the numbers before depressing the enter key. I noticed out of my window in *Aquarius* that there was a sea of debris riding along. Torn thermal blanket material and small kernels that looked like popcorn were floating nearby. There were shiny objects that had drifted further away that looked just like stars, so it was clear that it would have been impossible to do a fine alignment of our IMU by "sighting on" the stars. The completion of our power up and the final shut down of *Odyssey* went smoothly.

At one point, I noticed that Jack had left the dead *Odyssey* and was perched on top of the LM ascent engine cover. This was the first time that he had ever been in an operating LM. Jack's first utterance was, "Do you think this thing will make it?" The command module, in comparison with the LM, was built like a battleship to withstand the aerodynamic loads of a launch and entry. The LM, fighting a severe weight challenge, had the skin of the crew compartment walls chemically milled down to a thickness of .012 inches, which was the equivalent of three sheets of heavy-duty household aluminum foil.

Chris Kraft and several flight directors made the wise decision for us to loop around the Moon to get back home. In many presentations that Jim and I have made over the years, he stated that if nothing was done, we would have become permanent monuments, drifting in space forever. Years later, flight dynamics officer Chuck Diererich performed an analysis that had us missing Earth by 2,600 miles, looping back to the Moon, and running out of consumables if nothing was done to change our course.

With the change of the Mission Control team to Glynn Lunney, he requested that Bill Boone, as flight dynamics officer, and Tom Weichel, as retrofire officer, come up with a plan to get us on a free return trajectory that would land us somewhere on Earth. That somewhere was near Madagascar. So, at 61:29:43 elapsed mission time, Jim manually set the throttle to 10 percent for five seconds and then increased it 40 percent for the remainder of the burn. Attitude control was automated through the primary guidance and navigation system (PGNS) computer. Jim asked me to perform a consumable analysis because *Aquarius* was going to have to double its normal mission lifetime. I scribbled "grocery store" arithmetic at the bottom of a "burn" card. I often use a picture of that card when I talk to young school children to impress upon them the importance of learning arithmetic.

I figured that we could go down to a power level of 18 amps. A LM, in powered descent to land on the Moon, would have a power level of 65 amps. The unpowered coasting flight would use 32 amps. I calculated that the amp-hour capacity of our six LM batteries would supply us enough power to make it home. My figures also indicated that we would run out of water five hours before entry. But the saving grace was knowing that the Apollo 11 crew turned off the water valve before they left *Eagle*, as an engineering test, and the first critical system element failed in eight hours, so I figured that we were probably okay with our water-supply shortfall. I did not even bother to compute oxygen because, in addition to the *Aquarius* descent- and ascent-stage tanks, we had two full backpacks and two 6,000 psi OPS emergency containers available.

Meanwhile, back on Earth, we were a worldwide phenomenon. In West Germany, crowds gathered near newspaper offices to hear the latest. Radio stations in Paris broadcast special news bulletins and interviews with space experts. In

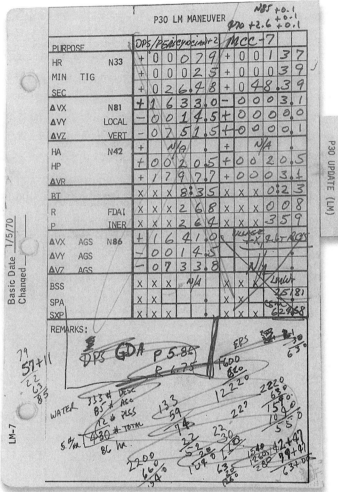

My handwritten calculations on my maneuver card to check if the available battery power and water would get us home. Courtesy of Fred Haise

England, the BBC covered the situation throughout the night. Australian TV continuously scrawled bulletins across the screen. In America, all three major networks produced live coverage of Apollo 13—a mission they initially didn't show much interest in.

A big error I made was failing to consider the lithium-hydroxide canisters as a consumable: The LM did not have an adequate supply onboard, even if we utilized the backpack canisters. The system spares, located on the descent-stage modular equipment stowage assembly, were unavailable, but even they would not have been sufficient. The workaround was to utilize the abundant supply of different size lithium-hydroxide cartridges stored in *Odyssey*. Ed Smylie, chief of the crew systems division, Jim Correale, Ed's deputy, and Jim LeBlanc, test director, plus other crew systems personnel, developed this modification, which was tested in a vacuum chamber at JSC, Building 7, where there was an actual LM environmental system. The step-by-step procedure that saved us came to be known as the mailbox and was made of a variety of low-tech items on the spacecraft, such as duct tape, stiff covers of checklist books, cellophane, and a couple of socks. I helped gather the material needed, and Jim and Jack performed the assembly.

As the backside of the Moon came into view, Jack and I took many pictures, because there was an abundance of film that otherwise would not have been used. Jim had already viewed this scenery for ten revolutions on his prior Apollo 8 Mission. He was feeling the disappointment, just as I was, of not landing on the Moon. Our path took us to an altitude just over 136 nautical miles above the back side of the lunar surface. Without appreciating it at the time, this put us into the *Guinness Book of Records* as humans reaching the furthest point from Earth. I also felt strange and lucky to have a firsthand look at this landscape that could only be seen from the vantage point of a spacecraft orbiting the Moon. The back side differed from the front side that we are familiar with from Earth—it was rougher looking, with craters upon craters. The primary colors were shades of gray and a few white areas that I assumed were newer craters. Just as when I witnessed the shrinking Earth, these novel aspects of the Moon made me question whether I was really seeing what I was seeing.

We shot a couple of photos of smooth areas that Farouk Al-Baz told us later in a debriefing were the best shots of those features, but I

One of the crew's photos of the Moon's back side, seen looking past the command module in the bright sunlight. Courtesy of NASA

suspected that Farouk was trying to make us feel better. Our photos included several of Tsiolkovsky and the Sea of Moscow. Tsiolkovsky was particularly striking—it was a smooth-looking black crater with a mountain in the center. Russia had circled the Moon first, so it earned naming rights.

I'm not much of a romantic. I'm more like Jack Webb on *Dragnet,* who wanted "Just the facts, ma'am," but the Earth was so much more beautiful than the beat up, drab Moon. The Earth is rich with color and from a certain angle it appears to have a halo. This is the Sun reflecting on the atmosphere surrounding the Earth. I can see why the environmental movement got a kick-start from those first photos of Earthrise on Apollo 8.

Jim told us to knock off our picture taking and get ready to join him for the PC+2—shorthand for pericynthion—the lowest altitude behind the Moon plus

Mission Control provided around-the-clock support. People on the ground ran tests and devised new procedures to get us home. Courtesy of NASA

two hours. We had received the maneuver pad of information to load into our computer, which would set off a four-minute, forty-second use of the *Aquarius* descent engine that would result in an 860-feet-per-second velocity change. This change would solve several issues. First, it would reduce our transit time to Earth by ten hours. Second, it placed our splashdown near the Samoan Islands. This was an excellent location because the ocean there is extremely deep. Depth was an issue, because the Atomic Energy Commission wanted the plutonium to be deposited as deep as possible. The LM descent stage had an RTG, which is what we called a SNAP 27 radioisotope thermoelectric generator. It had a radioactive plutonium fuel element that was the power supply for the Apollo lunar surface experiments package that we planned to implement. Third, the location placed us in the vicinity of the aircraft carrier *Iwo Jima* for recovery.

Our maneuver went as planned, with Jim setting the descent engine throttle to 10 percent for two seconds, then 40 percent for twenty-one seconds, and full throttle through shutdown. Several things were vying for execution at this time. One was to establish passive thermal control (PTC), which was a slow rotation of the spacecraft to ensure uniform thermal stress. In setting up the PTC, Jim manually established the designated attitude and made a yaw input with attitude thrusters. These actions caused the vehicle to tumble, which was not the desired outcome. Jim attempted this maneuver twice. Jim and Jack stayed on duty for more than twenty hours on the day of the explosion. I stayed up fifteen hours and only got five hours of sleep before rejoining them.

Eventually, we were able to work with Mission Control to power down, because it was important for us to preserve power. At this point, the vehicle was at about fifteen amps. MCC figured out several other actions that got us down to twelve amps. When I met with Colin Mackellar of Australia some years later, one power-saving innovation I found out about was NASA's request to reconfigure overnight their NASA Honeysuckle Creek Tracking Station's eighty-five-foot antenna near Canberra to support our mission. They also reconfigured a second antenna at Tidbinbilla and a third at Parkes. The eighty-five-foot antennae allowed us to shut off our S-Band power amplifier, saving another two amps. The upside of this was that we now had a consumable reserve, but the downside was the spacecraft was not designed to operate at that low power.

The heat from the minimal equipment running and from our bodies was not enough. Jim was worried about the water tank freezing, so he asked Jack and me to fill up some of the drink bags from the CSM food locker. Jack had earlier filled up eight, and together we filled twenty-two bags. We captured them in a drawstring storage bag. At one point, several drifted out and it took us some time to track them down. It was easy to lose something, even in our small spacecraft.

The environment became increasingly cold, eventually freezing the water tanks in *Odyssey*. There was also a water buildup on our instrument panel and other systems in the spacecraft because the water separator, having to operate well below its designed temperature specification, was unable to capture the moisture. We were not sure what the cabin temperature was, but I feel it had dropped to somewhere into the mid-30s in a day. We put on all our spare underwear, though unfortunately three pairs of cotton short-sleeve undershirts beneath our two-piece beta cloth garments did not help much. We considered donning our spacesuits, but we knew that without cooling air we would perspire. Jim and I had one advantage over Jack by donning our lunar boots. Jack had gotten even colder because his feet got wet from a leaky water gun near the ascent engine cover. The water guns were used to fill food and drink bags.

I looked at some photos of my family that I would have left on the lunar surface and wondered if I would ever see them again. I wondered how things were going at home. Thankfully, I knew that some of the wives and other astronauts would be at our home to support Mary. After the mission, I found out that my son Fred had stayed with our neighbor, Ed Smylie, because Fred and Ed's son were good friends in school. Ed had been instrumental in working out the carbon dioxide buildup problem in our spacecraft compartment. I am grateful to the many individuals who helped my family, especially Mr. Balch, the director of the then Mississippi Test Facility, which is now NASA Stennis Space Center. He knew my mother lived alone in Biloxi, so he dispatched test facility employees Henry Auter, Terry Malone, and Jack Rogers to stay with her.

At about ninety-seven hours mission time, I heard a thump. I also saw particles being emitted from the descent stage. The noise and dynamic effects were nowhere as severe as with the failure in our oxygen tank, but my thought was, what next? An electrical short in one cell of a descent battery had burst open and released the particles that floated around. Luckily, this had no further effect on the electric system.

At one point, Jack Lousma asked me, "How would you like to spend a week on an aircraft carrier," in reference to our post-splashdown. I responded, "If I can get on that aircraft carrier, I don't care how long I spend, Jack!"

Our trajectory was shallowing. Mission Control surmised that the small amount of thrust from the water sublimator was the culprit. For the third time, a midcourse correction was scheduled for 105:30:00 with the *Aquarius* descent engine. The added challenge was that we would have to execute this maneuver without the benefit of the PGNS computer. Jim attained the necessary attitude for the burn by utilizing a contingency procedure that he remembered from Apollo 8. He used the crewman optical alignment sight (COAS) mounted in his left window to view the cusps of the half Earth. The COAS projected a reticle image on the window, similar to a gun sight. He then pitched up very slowly. I viewed this maneuver through the alignment optical telescope that was normally used for star sighting. When I saw the sun appear in the sixty-degree upper view, I asked Jim to stop pitching. By then inserting a 400+5 entry into the abort-guidance computer's single register keyboard, a body axis align was set, to provide two steering needles or indicators on our eight-ball attitude indicator. The spacecraft was now at the correct burn attitude—which Jim and I were both involved in controlling. Jim took care of maintaining yaw attitude with the rotational hand controller, and controlled roll attitude by left-right inputs with the translation controller. I simultaneously controlled pitch attitude, with for-aft inputs of the TTCA. Jim started and stopped the descent engine, using 10 percent throttle for fifteen seconds. The attitude errors recorded were less than one degree, but the Hollywood movie years later made it seem like we were about to lose control.

Mission Control predicted that we would hear a puff from the pressure build up through the three descent engine firings as the relief valve cycled on the helium tank. The fact that we knew it was coming made it less jarring.

At one point, Mission Control told us to stop venting urine through the forward hatch because it was affecting tracking data. We used this alternate method because without *Odyssey* powering the overboard dump heater, the normal exit would freeze up. But with a shorter run through the forward hatch, there was no fear of freezing. As it turned out, Mission Control only wanted us to stop dumping for a few hours. We assumed they meant forever, so we subsequently faced a problem of where to store the urine. We utilized our urine collection device bags, backup Gemini bags, and the small canister where the backpacks were stowed. That canister was supposed to be used to store overflow water from the backpack servicing after our EVA, and it worked fine except for when the container was almost full, which caused a small bubble of urine to escape from the relief valve. Once again, the advanced technology of duct tape came into play to keep the valve closed.

Lazy me made the choice to leave a Gemini bag attached to my body for over a day to make my urine-transfer process easier, but this caused bacteria to build up, and I suffered a full-blown urinary tract infection. I felt horrible—it burned when I urinated, and I was quivering like a cold dog from fever.

Jim had been reserved through all of our collective trauma and lack of sleep in a cold, damp environment that was steadily worsening. After a while, he started to sound a little like the gruff Frank Borman, commander of Gemini 7 and Apollo 8. Jim spoke firmly when he told Mission Control to get us the *Odyssey* power-up procedure, because we needed to review it before executing it for the first time. I knew that the procedure had been worked by many, including some of our astronaut compatriots, proofing it through many runs in the CMS, but there is always concern with a first-run procedure. Finally, seventeen hours before entry, the procedure was read to Jack. He and I talked through the steps a number of times. It would have been nice to run through the procedure a dozen or so times in a simulator, but obviously we did not have that luxury.

There was a page in our LM checklist that listed the artifact and keepsake items that we planned to remove from *Aquarius* before its jettison. We planned to remove our hand and translation controllers, but because of our current situation, that would not be possible. So I started the removal operation. I took out all the inner wall netting material. The noise of the snaps being pulled loose bothered Jim a bit. I removed the overhead flood lights and the arm rests for the hand controllers, and I also cut the names off of our portable life support system backpacks as keepsakes. I put other loose items like the flight data file and storage items into a large drawstring storage bag.

At 133:24 mission time, we were instructed to power up *Aquarius*, which provided us with some warmth and rest time. When tracking indicated that our trajectory continued to be shallow, we performed another midcourse correction five hours away from our predicted entry interface. Jim wanted to perform the maneuver with the PGNS, but it had difficulty holding attitude, so we manually performed the maneuver for the twenty-two second burn, using four of our one hundred pound thrust attitude thrusters.

One of those several innovations was to use the *Aquarius* LM-7 batteries to recharge the *Odyssey* entry batteries. Had it not been for this, *Odyssey* might not have gotten through entry. Even with all of the increased computing memory and the advent of artificial intelligence over the years since, I think only a human mind could have come up with some of the ideas that got us home.

Four-and-a-half hours before entry, Jack and I had our cameras ready to photograph the separation of the service module. After the onboard countdown, Jack flipped the switch for separation, while Jim executed a short burst of the downward four-jet reaction control system (RCS) that was necessary for

After the service module separated, we could see one side panel had blown off and all of the damage that had taken place. Courtesy of NASA

clearance. Then he rotated the LM to get a good view of the service module, slowly drifting away. The amount of damage that I saw was hard to reconcile with the sound and motion experienced four days prior. A large side of the service module was missing, the high-gain antennae had been struck, and there was a dark streak on the side of the SPS engine bell.

While Jim stayed with *Aquarius* to maintain attitude and communications, Jack and I drifted through the tunnel to activate *Odyssey* at 140:10 mission time. I looked around in this cold, dead vehicle, with water covering the instrument panel, and hoped it would come to life to support our entry that was just two-and-a-half-hours away. Jack got some towels to wipe off the instrument panel, but he thought that this moisture was also on the wiring behind the panel. With that in mind, Jack told me that he would count down to the first step of closing all of panel five and eight's circuit breakers, on each side of the spacecraft. When he gave the go, we would each push in six circuit breakers and stop, then wait a short time, before counting off again to push in the next six. The few seconds in between would allow us time to smell burning wire insulation or to see smoke if anything shorted out. We continued until all of the breakers were closed and, thankfully, there were no problems.

Amazingly, we found that the abused, frozen *Odyssey* had not been damaged—all of the systems were fully operational. The activation checklist that the team on Earth created worked perfectly. The procedures also included forewarnings about caution and warning indications that could possibly arise so we would not think something else had gone wrong. Jack was able to follow up his course alignment with P52 star sightings to fine align the IMU, assuring a good attitude reference for jettisoning the LM and returning to Earth. After Jack verified his ability to control the attitude for the combined vehicles, Jim joined us in *Odyssey*.

Scientists and engineers from all over the world supported getting us home. To accommodate our s-band communications, the Australians made tracking station modifications. Another example was the Research Department at the University of Toronto, which had top experts in shock dynamics on its faculty. They were consulted to contend with the challenge of how to safely separate *Aquarius*. They agreed to study the problem and further enlisted their

Our view of *Aquarius* as it moved away from us before reentry. I wished
I could have brought it home with me. Courtesy of NASA

colleague at the University of Southern California, another shock dynamics
expert. Each of them performed independent analyses and passed their find-
ings on to NASA, which had completed its own analysis. Normally, the latches
would be opened while using the reaction control systems thrusters on the dead
service module to back away, but we needed a workaround, and the concept
was to pressurize the area in the tunnel to a level that would give a "kick" to
Aquarius at the time of its separation. However, there was a concern that too
much pressure could cause a leak in our tunnel hatch. In the final analysis,
the tunnel was vented to 2.2 psi before Jack lifted the covers and toggled the
CM-LM—FINAL SEP 1 and 2 switches. I had cameras ready to record the event
and was glad that I did not have my Hasselblad close to the window because

of the big jolt when *Aquarius* moved away. Earlier in the mission, I told Jack Lousma that I wished I could bring *Aquarius* home, but Capcom Joe Kerwin summed up everyone's feelings best: After the jettison, he said, "Farewell *Aquarius* and we thank you."

Twenty-seven years later, I gave a talk at the University of Toronto and had dinner with four of the six gentlemen in their Research Department who had come up with a solution. They gave me a copy of their brief study. I was impressed and amazed to hear the story of how they did not hesitate to help the Apollo program.

———

I laid in my couch awaiting entry interface, which was less than an hour away. Jim performed a secondary sextant star check that gave us great confidence on being in the right entry attitude. After the Moon occulted as further confirmation, Jack pitched down to entry attitude. As a spectator, I had a great view out of my window, looking in the direction behind our pathway into the atmosphere. The first thing that I noted when contacting the Earth's atmosphere was a white glow in the crew compartment. This was caused by the ionization of the air particles we first encountered as we entered at 36,210 feet per second, or about 25,000 miles per hour. A red glow began to develop as I felt the onset of acceleration pressing me into the couch. The red glow became a bright red trailing stream with an occasional small bright spot that appeared. I assumed it was a piece of the heat shield.

Jack had the entry being controlled automatically by the primary system computer, but, if needed, he could have taken over manually with the command module computer, or CMC. When the computer command caused the vehicle to roll for ranging, the fiery trail behind began to swirl. The acceleration or Gs kept building up to a peak of 5.56. This was the lowest of all the lunar entries. All the other missions were above six Gs. The red trail gradually turned to orange and then to a white, smoky trail. I heard the canister opening and could see the drogue chutes deploying to provide stabilization for *Odyssey*. On every mission, you know that you have it made when you see the main chutes deployed, starting at 10,000 feet. I braced myself for a hard

Me, Jim Lovell, and Jack Swigert exiting the helicopter. We were glad to be safely back on Earth. Courtesy of NASA

landing—on the previous Apollo 12 splashdown a camera bracket had broken free and Al Bean suffered a head injury when the camera fell—but luckily we had a soft splashdown. The chutes were released and *Odyssey* remained upright. Navy divers arrived to deploy a flotation ring around the capsule before they opened the hatch, letting out a cloud of frosty air. We exited into the life raft and then a Navy helicopter lowered a basket for each of us to board and be hoisted into the helicopter. NASA had our blue flight suits, sneakers, and Apollo 13 baseball caps there for us to wear for our grand entrance on the *Iwo Jima* aircraft carrier.

My legs felt a little wobbly as I climbed down the stairs from the helicopter to be greeted by Capt. Leland Kirkemo. The Navy chaplain spoke in a brief ceremony, and I headed off to sick bay, while my comrades enjoyed a celebration on the ship's hangar deck. A doctor gave me some aspirin for my fever and some orange juice.

We found out later that celebrations broke out around the world. Confetti and ticker tape were thrown out of windows in Manhattan, sirens sounded in Los Angeles, the bells in the National Cathedral rang for ten minutes, and

Mission Control broke out the cigars to celebrate our safe return. It was a well-deserved celebration. Courtesy of NASA

in Grand Central Station, people from all walks of life stopped and applauded. Even the Soviet Union helped by stopping broadcasts in the frequencies used by NASA and the recovery forces. As for myself, I was happy to hear that the reaction had been so positive, instead of a negative response for a failed flight. It was something we worried about on the way home. We felt there was a possibility that it would be a major setback to the Space Program.

Odyssey was hoisted aboard the carrier and briefly inspected. The water tanks were still frozen—the heat shield had done a great job! I thought our entry was another minor miracle. The electronic equipment in *Odyssey* had been exposed to an environment outside of its design specifications for four days, but it had come to life and given us a tie for the second most accurate splashdown of the program.

The next morning, we boarded the helicopter to American Samoa for a brief ceremony with the governor there. Then we flew to Hawaii.

President Nixon had flown to Houston to award Mission Control with the Presidential Medal of Freedom. Afterwards, Jim's wife Marilyn and my wife Mary joined the president on Air Force One for the flight to Hawaii. In a brief ceremony, we were awarded the Presidential Medal of Freedom. His words were inspiring:

Jack Swigert, me, and Jim Lovell next to President Richard Nixon as he awarded us the Presidential Medal of Freedom after the flight. Courtesy of NASA

I think I can truthfully say that never before in the history of man have more people watched together, prayed together or rejoiced together. This safe return is a triumph of the human spirit, the special qualities a man can rely on, and rely on all those things that machines cannot do.

Back in Houston the next day, I found out that I had a staph infection. The cure included two large injections a day in the buttocks for two weeks. Dee O'Hara was the appointed astronaut nurse who administered the shots. We had a code to remember the locations for the shots. The right cheek was "rise," or morning, and left cheek was "late," or evening. Dee came along on our postflight activity. When the crew appeared before Congressman Earl "Tiger" Teague's congressional committee, Dee and I had a hard time finding an isolated place in the US Capitol for me to pull down my pants for the shot.

—

Within a couple of months, Deke gave me my next job. It offered the oppor- tunity for another Moon flight. I was assigned as the backup commander for Apollo 16, which was slated to fly in early 1972. My crewmates were Bill Pogue, command module pilot, and Gerald "Jerry" Carr, LM pilot. Everyone on this mission was a member of the Original 19, and two of us were Marines. The prime crew was John Young, commander, Ken Mattingly, command module pilot, and Charlie Duke, LM pilot. My training to go to the Moon was interrupted by the many post-Apollo 13 media events in the first six months. The requests for the Apollo 13 crew made our week in the barrel seem like a cake walk.

In a *Life* magazine article, we wrote "Our Stories." I stated:

> we jettisoned *Aquarius*. I was sad in a way to see her go to her
> own separate fate—burning up in the atmosphere. I was proud
> of her. She was magnificent. Somewhat wistfully, I wished that
> *Aquarius* had a heatshield, too. I'd like to have that LM sitting
> right in my backyard.

Because of the staph infection, I still felt ill, so I missed a ticker tape parade in Chicago, but Jim and Jack, plus director of flight operations at MSC Sig Sjoberg and the four mission flight directors Gene Kranz, Glynn Lunney, Gerry Griffin, and Milt Windler rode in the parade. I did make the festivities in New York City, hosted by Mayor John Lindsay. On the receiving line in Rockefeller Center, we greeted several thousand people. A luncheon followed that included Grumman management and many New York dignitaries. Jack and I were awarded the Gold Medal of the City of New York. Jim was a previ- ous awardee of the Gold Medal after Apollo 8. We enjoyed dinner at Sardi's, where an impressive number of Broadway and Hollywood stars were there to meet us. I was still feeling down from my urinary tract infection and was disappointed that I missed the popular Broadway musical, *Hair*. There were some scenes degrading the American flag and Jim and Jack walked out. That got some press.

I felt particularly humbled that my hometown, Biloxi, Mississippi, staged a banquet and a parade to honor me. Our family rode in a convertible to greet the crowd, many of whom I recognized. At the banquet, I was given a watch

President Nixon with Jim Lovell, me, and Jack Swigert during a private
dinner with our wives and the president's staff. We ate in the White House
State Dining Room to celebrate the mission. Courtesy of the White House
Photographic Office

that was modified by a local jeweler, who replaced the Rolex insignia with the Biloxi lighthouse.

We were invited to the White House for a celebration of our mission on June 9. This was not a traditional state dinner; it was like a family dinner. The crew and our wives were hosted by President Nixon and his wife, Pat, in the State Dining Room, located a short distance from their living quarters. The President's immediate staff and their wives were also present, including H. R. "Bob" Haldeman, Donald Rumsfeld, and John Ehrlichman.

The First Lady carried more of the conversation than the president. We discussed the mission, but we also talked about everyday things. It was obvious to me that we were in the company of people who were very educated. It was hard for me to believe that these same people would be involved with the Watergate fiasco that later surfaced. The Nixons felt that we could use some relaxation, so after dinner, we flew to Camp David in Marine One. We spent the next couple of days there in a rustic cabin. Navy stewards prepared great meals for us. We tried out one hole of golf and Mary and I took short walks through a forest trail.

The two-week overseas goodwill mission labeled "Aquarius" commenced on October 1, 1970. We were accompanied by State Department personnel, as well as Gene Marianetti, of NASA Public Affairs. I was looking forward to the trip with Mary. I had been to Iceland before for geology training, but everything else on the itinerary was new. The first stop was at Reykjavik, the capital of Iceland. We attended a social event at the US Embassy and visited the Þingvellir National Park. There was a striking wall of rock there, where the founding Althing National Parliament met to hold their sessions from 930 AD. I imagined the leadership all clad in Nordic clothing, standing on the cliff, looking down on us.

From there, we headed to Switzerland, stopping in Zurich, Bern, and Lucerne. In Bern, we were scheduled to appear at the Federal Assembly. We arrived early, so I separated from Jim and Jack to visit a small gift shop just down the street. I wanted to purchase a small knight in armor to bring home. As I walked back to the Parliament Building, a man getting off a bus asked me if I was Fred Haise. Wow, I felt I had come a long way from small-town

Biloxi, Mississippi—a stranger recognized me and called me by name in Bern, Switzerland. I was really floored when he introduced himself as Hans-Peter Tschudi, the president of Switzerland. This encounter told me a lot about Switzerland—their leader could roam without a security escort.

I enjoyed the visit to the Swiss Museum of Transportation in Lucerne. I felt right at home seeing all the airplanes, trains, and automobiles. We made a side trip to Lake Constance, which is bordered by Switzerland, Germany, and Austria. We made a press appearance at the Twenty-First Astronautical Congress. Several cosmonauts were in attendance. One of them had just returned from a mission on the Soviet Soyuz 9.

Next, we journeyed to Athens, Greece. This was following the overthrow of King Constantine and the Greek government on April 21, 1967, by a group of military officers. The government that followed is known as the Regime of the Colonels. Our visit represented an early US government recognition of the regime, with Prime Minister George Papadopoulos at the head. What struck me most as we drove around the city was the appearance of so many armed soldiers and even a few Army tanks. We had an official, brief visit with the prime minister, and I felt that I had to be at attention the whole time. Jim presented him with an American flag and patch from our mission. Activities for our official stops were planned by our friend, Gene Marianetti, whom we had all worked with previously on media events. We also stopped at the University of Athens and the Tomb of the Unknown Soldier. As tourists, we visited the Acropolis and Parthenon. With all that geology training under my belt, I couldn't help but relate to our guides discussing the age of those historic sites. The rocks used for constructing them had been around for millions of years prior. The food in Greece was great, but everything seemed to be cooked in olive oil, even scrambled eggs. Even bread was accompanied by a small dish of olive oil.

Our schedule included a few days off to rest from the gauntlet of events. We journeyed to Crete. Jim, Jack, and I had a chance to dive in shallow water to look at some old artifacts that were easily seen in the clear water. A brief visit to an archaeological site from the Minoan civilization was enjoyable. I was fascinated by their water transport elements for showers, drinking, and waste. The pipes had the same swaged ends that you would find today in the United States. Our advance press must have been thorough, because practically everywhere

I went, a group of children followed me. I felt like the Pied Piper of Hamelin.

Our next destination was the State of Malta in the center of the Mediterranean Sea. We enjoyed the equivalent of a ticker tape parade down Kingsway Street in the capital city of Valletta. The route was crowded with people, including many perched on balconies along the way. A formal dinner was held with government leaders and prominent citizens. As with most events of this type, I found the atmosphere to be stiff—it made me wish I was somewhere else. Toasts were made and, fortunately, Jim was stuck with the honor. I am not good with toasts if they are longer than "cheers," "*salud*," "*skoal*," "*prost*," or "bottoms up."

The last country we visited was Ireland. We met President Eamon de Valera and Prime Minister Jack Lynch at the president's residence. There were lots of cheering people along the way for our motorcade. We stopped at St. Patrick's Cathedral in Dublin. I felt indebted because so many people across the world held around-the-clock prayer vigils for our safe return. While on the mission, it never occurred to me that so many people cared about us.

The highlight of our visit was the ride on an ancient train through Ireland down to Cork on the south shore. We made stops at towns along the way and greeted people from the train car platform. It reminded me of how politicians had campaigned in the days of yore in the US. The Irish peoples' enthusiasm was impressive. At one stop, in our Cork motorcade, Jack Swigert climbed out of the car window to try to wave the people back because it looked like some, including children, might be crushed. We left Cork to spend a night at the Dromoland Castle near Shannon Airport. The castle had been revamped from those damp, cold, dark, olden days. It was quite comfortable and we enjoyed our five-star accommodations. The next morning, we headed home.

—

CHAPTER 11

BACK TRAINING FOR THE MOON

Much of the training for Apollo 16 was old hat. There were a few new things, including the lunar buggy and some of the lunar Apollo Lunar Surface Experiments Package (ALSEP) experiments. The LM-11 *Orion* had grown to provide the consumables to support a longer stay on the lunar surface for three EVAs. In the training lineup, I was behind Apollo 14 and 15, so it was a challenge to find simulator time again. With the extended two-year training cycle ahead, John Young decided that we should take advantage of the time for more geology field trips. With John's okay, I signed up for a second job as a CapCom for Apollo 14. I joined Gerry Griffin's gold team because it would be on duty at the time of their planned second lunar surface EVA. The EVA traverse on the flank of Cone Crater was what I had trained for on Apollo 13. I thought my experience might be of some help to Al and Ed. I enjoyed working through all the integrated simulations as a member of the Mission Control team, meeting the challenges thrown at us by that devious Sim-Sup gang.

Apollo 14's crew was Al Shepard and two members of my Original 19 group, Stuart "Stu" Roosa and Ed Mitchell. The Apollo 14 Mission launched on January 31, 1971, after a forty-minute weather delay. All of the spacecraft modifications recommended by the Apollo 13 Accident Board were accomplished within nine months. A potential mission-abort problem showed up as Stu

Roosa attempted to dock with the LM-8 *Antares*. The latches would not fire to lock the two vehicles together. After six unsuccessful docking attempts, over nearly two hours, Stu tried a different procedure that required good piloting techniques. Stu retracted the docking probe all the way and eased *Kitty Hawk*, the CSM, into contact with the LM and held the vehicles together using plus-x thrusters. Much to the crew and Mission Control's relief, the latches fired and locked the spacecraft together.

The flight went well, but the next challenge arose as Al and Ed separated *Antares* from the CSM and were getting ready to land on the Moon. Once again, I experienced a sick feeling, fearing that the mission would be aborted because an abort-discrete warning showed up on the computer. This would result in the computer being switched from P63, for landing and braking, to the P70 program, for an abort using the descent engine. Don Eyles at the MIT Draper Labs saved the day. He was responsible for the LM PGNS system and figured out a workaround that Ed Mitchell typed into the DSKY keyboard, but the drama was not over for Apollo 14. As Al and Ed approached pitch over, the landing radar had not locked up, and, if it didn't, this would call for an abort. I am sure this raised the adrenaline level a bit with Al and Ed thinking, "What next?" Flight director Gerry Griffin told the CapCom to have Ed cycle the landing-radar circuit breaker. That action was another one that saved the mission, resulting in a radar lock on to feed the computer with precise altitude data.

As it turned out, Al executed the most accurate touchdown of all the lunar landings, relative to the planned location. The Fra Mauro site was chosen because the formation was thought to contain ejecta from the massive meteorite impact that formed the Imbrium Basin to the north. Many of the rock samples the crew returned were breccia-type rocks that are formed through the pressure and heat of a meteorite impact. Some of the basalt fragments found in the rocks were more than four billion years old. They were older than any samples returned from all the other Apollo landing sites.

The primary goal of their geology traverse on the second EVA was to sample the slope of the 1,000-foot-wide Cone Crater. Jim Lovell, when he first studied the lunar maps of Fra Maura, gave the crater that name. As we all had learned in our training on impact features, the ejecta provides samples from the deepest material all the way to the surface material near the crater rim.

Moving up the slope turned out to be quite a workload because of the modular equipment transporter, which was supposed to have eased their workload, but they had to drag it along or carry it. They were breathing heavily and running low on their oxygen supply, so the flight director told me to request that they cut short their EVA before reaching the rim. It took a couple of requests, and my good friend Ed called me a fink. At the time, I wasn't sure what that meant, but I knew it wasn't nice. I looked it up later and found that it meant an "unpleasant or contemptible person." In their shoes, or should I say lunar boots, I would have felt the same way, but I did not want them to die on the Moon from lack of oxygen. The nine-day mission ended with a splashdown in the South Pacific Ocean on February 9, 1971.

Jerry Carr and I worked together on the first three of seventeen Apollo 16 geology field exercises. From July through October of 1970, we trained at the San Juan Mountains in New Mexico; Medicine Hat in Alberta, Canada; and Colorado Plateau.

I received a bit of bad news when NASA announced that Apollo 18 and 19 were canceled. I knew I had lost any possibility for my second trip to the Moon. And, since they had not flown yet, and not many open seats remained, Deke assigned Jerry and Bill to the last Skylab Mission. Stu Roosa and Ed Mitchell were assigned to join me on what turned out to be a deadhead backup crew assignment.

One of the more interesting sites was at Sudbury in Ontario, Canada. There was a sixty-mile crater that was deemed to have been the result of a meteor impact some two billion years before. It was of particular interest because the area offered great examples of shatter cones, a byproduct of an impact event. This environment is like the lunar environment, which has no end of meteorite-impact features.

Ed and I worked together through the last five exercises at the Rio Grande Canyon near Taos, New Mexico; a Nevada nuclear test site, Coco Hill, California; Hawaii; and Boulder City, Nevada. The last geology field trip began on February 17, two months before launch.

There was a 1-G trainer for the lunar rover that was used on several geology field trips. We also went through deployment and setup of the rover, from the LM trainer vehicle, in the KSC Training Facility. The rover had four-wheel

Dr. Lee Silver pointing out geological features to Charlie Duke, me, Ed Mitchell, and John Young while on a training mission for Apollo 16. Courtesy of NASA

drive and was equipped with a hand controller, rather than a steering wheel. The training rover had automobile tires with the treads shaved off, rather than the wire wheels on the actual rover vehicle. The frame and structure were much stronger on our training rover than the real rover, which would have collapsed from our weight in 1-G on Earth. It also had a simple, but effective, navigation system. Unfortunately, I did not get to drive the vehicle on the Moon, but John and Charlie covered 16.6 miles on their three EVAs. John also had the fun of conducting an impressive lunar "Grand Prix" test of the rover through various speed and turning maneuvers that can be found in a video online.

Apollo 16 launched on April 16, 1972, two years after Apollo 13. As a good backup, I followed the mission from a position by the CapCom in Mission Control. This was the third time I observed others doing what I had been trained to do. There were a few minor anomalies early in the flight, as with almost every other mission, but a potential lunar landing abort came up just after

John Young, Gene Cernan, me, Charlie Duke, Tony England, Gordo Fullerton, and Don Peterson as we completed an evaluation on the lunar rover in 1971. Courtesy of NASA

separation of the LM *Orion* from the CSM *Casper*. An oscillation was noted in the CSM SPS engine's secondary yaw gimbal actuator. This required John to make a brute-force rendezvous to keep the vehicles within close proximity for three revolutions. After a discussion and test by the contractor, Jim McDivitt, the Apollo program, director gave the go-ahead for powered-descent initiation more than five hours later than planned. With the orbital position changed relative to the landing site, retargeting was required. John landed at the Descartes site with one hundred seconds of hover fuel remaining.

A problem also occurred on the first EVA while setting up the ALSEP experiment array, which had various experiments laid out around the Central Station. John caught his leg in the cable to the Heat Flow Experiment and tore the cable. Don Arabian, manager of the Program Operations Office, came up with the idea to use a lunar rock as sandpaper to expose the wire to repair it with duct tape. In my three backups, this was the only time I did something that solved a real problem. I suited up and went to work on that wire with the

Moon rock. Due to time required, John did not get to attempt the repair on the Moon.

On the three EVAs, John and Charlie collected 211 pounds of rock samples. Their traverses took them up steep slopes. At some of the stops they parked the rover in a shallow crater to assure that it wouldn't roll off, because the rover had no emergency brake. The return to Earth was normal.

———

After each mission, NASA threw a Pin Party. This was a banquet for all of the astronauts. Al Shepard would award a gold astronaut pin to those who had flown a mission for the first time. When one first became an astronaut, one received a silver pin. After the dinner, backup and support crew members roasted the crew that had flown. For Apollo 8 we got a little theatrical. Some of us popped out in various costumes to spout appropriate one liners. We got this from the popular TV show *Rowan & Martin's Laugh-In*.

For Apollo 11, we used a different approach. Jim Lovell was our master of ceremonies. Some of us formed a quartet to sing satirical lyrics, although the voices left much to be desired. I played Neil Armstrong and had money coming out of every pocket of my sport coat. I spent a lot of time learning to mimic Neil's voice to deliver his Wapakoneta, Ohio, accent.

For the Apollo 16 KSC lunar field EVA exercises, I had one pack of film for a battery-powered 16mm camera. That film was used to tape a segment for the Pin Party video that was planned for post-mission. From my limited video production experience, I learned of the challenges that movie makers experience when I was missing a prop as we were ready to film. So I learned to write a script, including the prop list for each scene. Some scenes had to be shot after the mission to capture some events, like when John broke the experiment cable or had stomach problems. The latter was caused by the crew being encouraged to drink copious amounts of orange juice to offset the heart arrhythmias that were experienced on Apollo 15. Using a colorful expletive on a hot mic, John expressed to the world that he would never drink orange juice again. The state of Florida did not hold that against him: State Road 423 in Orlando, Florida, is named the John Young Parkway.

After four years of intensive training for four lunar missions, I thought I might have some time with the family for relaxation. But within a couple of weeks of the Apollo 16 splashdown, I was assigned to join 416 personnel, including twenty members of the Air Force, to evaluate proposals from Rockwell, Grumman, McDonnell Douglas, and Lockheed to build the space shuttle. Our evaluations and meetings took place seven days a week, but at least I was home with my family at night.

I joined the Integration Committee led by Bass Redd. We assessed design and personal accommodations, flight and system analysis, aerodynamics and thermodynamics, and payload accommodations. Our group included twenty-five people from disciplines that included engineering and development, flight crew operations, data systems, and analysis directorates. Bass was the engineering analysis division assistant chief for the space shuttle. I had not worked with him before, but felt he did a great job of capturing the salient details from the diverse participants. I also came to know that Bass was a deeply religious person.

I first dug into the short executive summary and the technical volumes to get a picture of the space shuttle orbiter—a vehicle that I hoped to fly someday. All of the proposed configurations for the orbiter were winged vehicles except Lockheed's that proposed a lifting body. From my previous experience with the lifting body program during my Flight Research Center days, that configuration was good to handle entry heating. On the downside, it would not be as good a glider with its lower lift-to-drag. I was surprised to see that the orbiters would have jet engines that were deployable late in entry. I suspected that this was because there was insufficient knowledge of the hardware, such as the IMU, to be confident that the navigation capability allowed for landing on a runway, rather than in the wide expanse of an ocean. All of the proposed vehicles were equipped with a solid strap-on rocket to provide abort capability right off the launch pad. The materials proposed for the orbiter were titanium, a nickel-chromium alloy known as Inconel X, and other metals. The Rockwell Corporation was the winning bidder with a score of 603, versus the Grumman Corporation's 597.

—

During this era, I began to think ahead about what I might do when I was too old or not otherwise physically fit enough to fly. If I couldn't fly, the next best thing would be to have a hand in the creation of new flying machines. I thought program management was a fitting career, but I had no business experience—except on a minuscule scale as a newspaper boy. I talked to my boss, Al Shepard, and Chris Kraft, director of JSC, about my future goal, and before I knew it, I was sent to the Harvard Business School Program for Management Development Class 24, along with Jim Bone, who was deputy manager of Shuttle Resources and Schedules Integration at the time. So I was off to Boston, leaving my family once again for four months in 1972.

The Program for Management Development was a four-month pressure cooker course. Classes were held six days a week. On the seventh day, we were expected to attend scheduled social mixers. There were 156 people in our class from all over the world, though none from China or the Soviet Union. Our class was divided into two groups for classroom lectures. The classrooms had an upward tiered seating arrangement with the instructor at the bottom. Halfway through the curriculum the groups switched places. The intent was to allow for mixing within our class to experience the many cultures in attendance.

We lived in a dormitory, divided into groups of four. My roommates were Herbert Hoffman, general manager with the Loews Corporation; David Rowe-Beddoe, director of Thomas De La Roe Company; and Rodrigo Peluffo, director of administration and finance with LePitit SAQIC in Argentina. I was impressed that my English roommate, David, had been one of the pacers when Roger Bannister first broke the four-minute mile. Sadly, Rodrigo was killed a few years later on March 3, 1974, in a plane crash on takeoff from Orly Airport in Paris. The cause was a rear cargo door opening on the DC-10. I reviewed the hatch design, as well as other aircraft accidents involving systems that might be applicable to shuttle, in my later Orbiter Project Office role.

My roommates decided that I should run for class president. I was not interested, but I figured it would not be a distraction because the class officers had no real responsibility. My new friends played the famous-astronaut card to get me elected. The one distraction that came up was with a woman from Africa who was actually a princess. The issue was she cooked in her room beyond dinner time, preparing food that was heavily scented with the spices she used back home. Some of the other dorm residents in that part of the building complained to Harvard management and, when that failed, to their class president. Cooking in a dormitory room was forbidden, but Harvard didn't want to cause a fuss with their royal student. So, in order to prevent an international incident, I met with her to negotiate a solution. She explained to me that she did not like the local food. I explained that the cooker was illegal, per university rules, but told her that she could continue to cook as long as she finished near the end of normal dinner time. I thought her cooking smelled good.

The curriculum included a total of 218 classes covering marketing, manufacturing, finance, planning and control, decision analysis, organizational problems, human behavior, labor relations, economic environment, and business simulation. Finance included some accounting basics that were foreign to me. The case-study method of teaching was used for each class. Overall, the approach was effective and interesting, because the cases were about real companies that had problems related to the subjects covered in the classes. However, the entire program was directed toward private business and large corporations rather than government. Our wives joined us in Boston for a few classes at the end of the course, and we all enjoyed a great social gathering to celebrate.

After arriving home in Houston, I met with Aaron Cohen, space shuttle orbiter project manager, to discuss my new job as his technical assistant. The description listed a broad scope of responsibilities that covered all aspects of the orbiter design and development. I requested that one line be added to allow me to maintain flying currency.

Each morning at 7:30, I joined Aaron for reviews by our subsystem managers. There was a subsystem manager assigned to each of the systems in the orbiter, so it took almost two weeks to cycle through all the subsystem reviews. Every Saturday, we held a teleconference with Rockwell management to review the program status. Depending on the open items or technical issues, sometimes a subsequent review was scheduled at Rockwell or the pertinent subcontractor's facility. One of my job responsibilities was serving as a member of a board that reviewed all orbiter engineering change proposals. I enjoyed working for Aaron. He was an excellent mentor and manager and had the trait of other great NASA managers, such as Bob Gilruth and George Low, who were great listeners. I had observed Bob, seemingly almost asleep, suddenly ask a presenter a question that often made them wish they had done more homework. I witnessed George Low sit through a full day of engineering change reviews, particularly in the days when there were many changes due to the Apollo 1 tragedy. George, as with Bob and Aaron, let everyone share their expert opinions and facts before making a decision. And many difficult decisions had to be made, despite the fact that sometimes, even with all the possible data at hand, the picture was still gray, not black or white.

I found out quickly that managing a program is a continual struggle to balance the budget available against program content and schedule. Almost immediately, we faced the challenge of orbiter cost reductions due to Congressional underfunding. Some of the easy solutions proposed were to delete the deployable jet engines and the launch-escape solid rockets. Additionally, a second orbiter vehicle that was proposed to support the Approach and Landing Test (ALT) Program was deleted. There were many other cost-saving decisions that Aaron made, including the deletion of thermal-vacuum and vibro-acoustic test articles. The Structures and Mechanics Division staff, including Tom Moser, Phil Glynn, and Tom Modlin, seemed to raise Aaron's ire. Phil caused him to break his pipe on the desk at one meeting. The meetings were heated because the deletions to save money potentially meant risking the vehicle and the crew's safety. I was concerned about the plan to delete the orbiter aft section vibro-acoustic test article because that's where the highest acoustic loads occurred, which could result in structural failure. I requested Aaron's permission to schedule a follow-up review of that specific deletion with Chris Kraft, but Chris agreed with Aaron on the deletion.

In addition to the cost reduction efforts, there were weight-reduction concerns. Every new flying machine has faced that challenge in development. One consideration for weight reduction was in the area of vehicle wiring. The proposal was to connect all system components to a computer. The deletion of cables and pulleys that are directly connected to the flight controls triggered many concerns for us, as pilots. *You mean we have to depend solely on a computer and software to move our control surfaces, fire attitude thrusters, or gimbal engines?* This whole idea took some getting used to in the Astronaut Office, and adding insult to injury, the plan being hatched depended on four computers to operate in harmony. Developing workable software for this architecture turned out to be a major challenge.

Additionally, the crew cabin was blunted and the wing area was reduced. There were other proposed deletions that would have caused the vehicle to become a less-capable glider: Therefore, the heat load would increase during entry. The peak heating that was predicted was above the capability of the exotic metals proposed. Fortunately, Lockheed had developed the LI-9000 silica material that was improved by the coating work done at NASA Ames Research Center, which saved the day by using lightweight thermal tiles. Over 24,000 tiles blanketed the orbiter. For entry temperatures over 2,300 degrees Fahrenheit, a carbon-carbon or RCC material was utilized for the nose cap and the wing leading edge.

I got a phone call from John Young because he was concerned about Rockwell's plans to film design review meetings. He wanted to make sure that any release of crew photos would be cleared through NASA's Public Affairs Office. John also thought that the Project Office should resume the widows-and-orphans talks at contractor facilities. These talks were given during the early stages of the Apollo program by an astronaut to the workers on the floor—to all shifts if possible. The intent was to instill in the workers a sense of the importance of their work and to heighten their conscience to do the job correctly the first time. The presentations would begin with a big picture of the mission and then focus on the specific component the employees were working on. Through this process, they were told that their job performance affected mission success and crew safety. I planned to fill this role when I could, but when an astronaut happened to be near a contractor facility while on a public relations assignment I coordinated with the Astronaut Office. It made sense for them to give a widows-and-orphans pep talk.

My assignments included reviewing specific activities or reports, and giving Aaron my synopsis. I also independently studied unresolved space shuttle issues as they related to the orbiter, and discussed them with Ron Kubicki, Engineering Office manager. There were forty-one open items or issues, so we had many things to figure out. I prepared quick-look summaries on the life-sciences-waste-management system and a Spacelab Level I Document with respect to orbiter impact. Aaron wanted my assessment of the requirement for an unmanned flight and of the study being conducted by Charlie Duke, who had done a great job of describing each mission phase from launch to landing, and how a crew would improve the likelihood of success. My report discussed the potential impacts that the development of unmanned capability would have on the design and schedule.

—

On August 23, 1974, as a thirty-nine-year-old pilot, I encountered another one of those sharp turns in the road after I joined a group then known as the Confederate Air Force and now as the Commemorative Air Force. They staged air shows with authentic World War II aircraft or replica look-alikes, and the opening act of each show was a simulation of the attack on Pearl Harbor. A number of planes had been modified by Twentieth Century Fox for the movie, *Tora, Tora, Tora*. The *Canadian Harvard* had been modified to look like a Japanese Zero aircraft. Similarly, Vultee Vibrator BT-13 Valiants were modified to look like Japanese Val dive bombers and Kate torpedo bombers. I flew the Val, diving through the melee of aircraft that included Zeros, Kates, a B-17, and a single P-40 fighter that Joe Engle often flew. A slew of pyrotechnics were ignited to create booming sounds and lots of smoke. On that fateful August day, I was ferrying the Val from the crop-duster field near Angleton, Texas, to Galveston Scholes Field, to have it cleaned for an upcoming air show at Dallas, Texas.

Ted Mendenhall, one of our shuttle Gulfstream instructor pilots, was flying lead on the short flight. On the landing approach, I found myself too close to Ted's Zero aircraft. I was concerned about overrunning him, so I added power to execute a go-around. At about 300-feet altitude, headed south toward the Gulf of Mexico, the engine started sputtering and cutting out. This was clearly one of those *never panic early* moments. So, I switched fuel tanks, even though I knew both had fuel, and rapidly pumped the wobble pump handle to increase fuel pressure. With that technique I got the engine to run for a bit, but then it started sputtering again. I tried to quickly figure out the best option. I did not want to land in the Gulf because the water would be shallow.

The Val also had a fixed landing gear, and I feared that it would flip over when impacting the water, leaving me upside down with the aircraft on top of me and no way to swim clear.

As I milked it around through 90 degrees of turn, there was a field with cows and a canopy of trees. I thought that I could land with the nose between two trees and only harm a few of the cows. With the engine alternately running and quitting, I worked my way around another 90 degrees and decided to land downwind on the west side of the airport. A compounding problem was caused by the metal plates that were installed to prevent the landing-gear oleos from stroking and damaging the ornamental parts that were installed to look like the Japanese Val. As a result, only the rubber tires provided shock absorption for landings. Shortly after touching down hard on the uneven ground, one landing gear failed. This resulted in a wing tip crashing into the ground, which caused the plane to cartwheel. When the plane came to a stop, I was upside down, with the canopy closed and blue flames coming out by my feet and the rudder pedals. I unlatched my harness and turned to kick a hole in the canopy. At this point, my rear end was in the flames, and the flames ignited the arms and legs of my cotton World War II flight suit. When I escaped, I rolled on the ground to extinguish my burning suit.

A man and his son showed up with a small rug to further dampen the flames that had enveloped me. They saw the crash from their backyard and had already called for an ambulance. They helped me get through the airport perimeter fence to sit on their porch, and the man's wife brought out a bowl of ice that she thought would soothe my burns. Then my vision started to white out, and I was concerned about my future flying career because I thought I had burned my eyes. I found out later that this was a manifestation of going into shock.

The ambulance arrived with two sheriff cars for an escort. Some of the sheriff personnel were members of the Confederate Air Force and wanted to make sure I got to the hospital as quickly as possible. My boots were cut off in the ambulance, but no other treatment was given.

I would describe arriving at the University of Texas Galveston Hospital as entering the door to Hell. The first procedure I underwent was climbing into a tank of water to help remove the cloth flight suit that had burned on my body. The pain was severe. It seemed to worsen when I was told I could not

The Vultee Vibrator BT-13 Val dive bomber flipped over after doing a cartwheel.
I managed to get out, but with second- and third-degree burns.
Courtesy of Fred Haise

receive any pain medicine until X-rays cleared me of internal injuries. After what seemed like forever, I was taken to a hospital room and given some pain killers. When I met with the doctor he said it appeared that I had 65 percent burn coverage, with second-degree or third-degree burns. He also told me that I would be listed as critical for a couple of weeks while he observed my condition and monitored my respiratory system to see if it had been burned from the hot gases.

I was blessed to have twenty-four-hour support from family and friends. Mary, her sister Susan, and visiting Biloxi family stood first shift. Second shift was covered by my buddies from testing the LM, John Presnell, Henry Gawrylowicz, Dave Ballard, and Hal Taylor. The third shift was manned by several from the Original 19, particularly Charlie Duke, Ken Mattingly, and Paul Weitz.

After ten days, they took me off the critical list and told me that I could look forward to living for a while longer. I asked for a meeting with my medical team, including Dr. Dwayne Larson from the Shriner Burn Institute. Dr. Larson was going to perform grafting on the third-degree burn areas. The Shriner doctors supported the University of Texas adult burn ward and trained UT residents who were to become plastic surgeons.

I told my doctors that my goal was to return to flight status. In discussing the upcoming treatment, they only came up with one possible impediment—I would require grafts around my legs. After grafting, there could be no pressure on the grafts for five days. The normal practice would have been to put a pin through my ankles to allow my legs to be suspended in the air for the five days. They thought it possible for a cavity to be left in the bone structure, which could cause a problem with pressure differential when I flew, considering the altitude changes. NASA Crew Systems came to the rescue with a workaround—they put Velcro on the bottom of my slippers and added Velcro on a wooden board attached to the foot of my bed. This allowed me to simply plant my feet onto the board to keep my legs airborne after grafting.

A daily routine was followed to prepare me for grafting—it was torture. One hour before the bath in a Hubbard tank, which was filled with diluted Clorox water, I received a morphine shot. I was worried about getting addicted, so I asked for a pain pill rather than the shot thereafter. The Clorox prevented infection from developing in my large open wounds. When my appointment approached, a couple of large handlers arrived. Keeping my body rigid, one handler gripped my shoulders and the other my feet. They lifted me onto a gurney positioned beside the bed. At the bathing room, I was hoisted into the air and moved over to be lowered into the tank. There was a timer on the wall with the dial showing seventeen minutes. The tank was shaped like a violin to allow two nurses to position themselves in the indented areas. This allowed them to remove the loose, dead skin with tweezers and scissors. The pain level seemed to grow during those seventeen minutes, and I just kept reminding myself that Marines were tough. The crane lifted me back into the air and a nurse had a hose ready to rinse me off with the water temperature set to 98.6. Then, I was placed back onto the gurney. At this point, I was freezing, literally shaking like a dog. To help with this, nurses installed hoops on a gurney to support warmed sheets that they draped over me for the ride back to the room in my mini covered wagon.

But the daily ritual was not over yet. There were large heaters set up on each side of the bed to keep me warm. My nurse was waiting and continued the burn debridement on the third-degree areas with tweezers and scissors. Because of the wait to determine what areas were second and what were third, my body had developed a covering called eschar skin over the open third-degree burns. It was white and milky looking, but would have to be removed through debridement before grafting. During this several hour process, my nurse often had the TV program *Soul Train* playing. For the next several years, whenever I heard the *Soul Train* theme music, I immediately remembered those unpleasant days. The nurse completed the process by wrapping my limbs in gauze, impregnated with silver sulfadiazine. Then she topped it off with an Ace bandage wrapping.

My wife brought an electric razor to keep my whiskers trimmed, which made me notice that the rooms did not have mirrors. It was not considered good for morale to see yourself in a mirror, particularly if one had facial burns. For a couple of weeks, I had one red ear and a red area on my neck from first-degree burns. Patients with burns all around their torso would have to be rotated to prevent bedsores. They had water beds with the temperature kept at body temperature, but it was clear from the screams that I heard when someone was being turned, during the day or night, how some patients were so badly burned.

I quickly gained an appetite and was told to consume lots of calories. A hospital nutritionist came to visit me when she heard about my accident. She was from Gulfport, Mississippi, and we bonded on that basis because my hometown of Biloxi was only twelve miles away from hers. She began preparing rich milkshakes for me every day. Because of the fluid loss through my open wounds, I also was given a small carton of milk to drink every couple of hours.

I had lots of visitors over the eleven weeks I spent in the hospital. To keep me abreast of what was going on, my Project Office colleagues brought copies of memos or presentations covering orbiter development. Dr. Larson also brought several other physicians by for a visit. One day, in the midst of my Clorox bath, he brought an elderly doctor to see me. He had white hair and a British accent. He told me that he had worked in a large hospital outside of London, with five hundred beds. Many of the beds were occupied by Royal Air Force (RAF) pilots who flew the *Spitfire*, which had a fuel tank that was not self-sealing right in front of the cockpit. When the German ME-109 shot bullets into that tank, the RAF pilots would find themselves aflame, caused by burning fuel in the cockpit. To make matters even worse, the British pilots often bailed out in the cold English Channel to await rescue. Many times, they would suffer facial burns. Somehow this made me feel that I wasn't so bad off after all.

———

My friend, NASA pilot Ed Rainey, had an accident caused by a landing gear problem and ended up landing on the belly of the aircraft at Ellington Field. As a joke, I had an official-looking certificate created and framed that lauded his "test piloting" skills in testing the effects of a concrete runway on an aluminum structure. When he came by to visit me in the burn unit one day, he had his own idea of a joke. It was a large card carrying the prominent statement, "Haise, you can't even commit hari-kari successfully!"

I was told that I should walk as much as I could. During the first week, I needed handlers to get down the long hallway and back to the room, but as I got stronger, I continued on my own and began to explore other areas of the hospital. I came across a long room where several people were just lying in their beds, looking like they were asleep. I was told they were in comas, and I was once again reminded that my situation wasn't so bad. Each time I returned to the room, the Ace bandages covering my legs were red with blood. During my hospital stay, I received seventeen blood transfusions.

The day finally came for my grafting surgery. Dr. Larson operated on me for five hours, and I was pretty much out of it through part of the following day. The challenge of urinating with my legs hoisted was easily solved. I asked a NASA suit tech to bring a urine-collection device to use. Dr. Larson saw the advantage of that device and worked with the manufacturer to develop a simpler disposable version to accommodate the needs of the young boys in the Shriner Hospital next door. My grafts were healing well. Within five days, I was once again taken to the tank bath—this time with plain water. The wrappings around my legs were carefully cut loose and I happily noted no pain. One small area of the grafts had come off, but Dr. Larson replaced it with some skin he had saved. Most of the skin for grafting had come off of my chest, and a crusty scab layer developed. Charlie Duke came by to visit and, using his best bedside manner, told me I looked like an armadillo.

After another week, I was cleared to leave the hospital. Surprisingly, as miserable as my experience had been, I felt a little sad to leave my home away from home. I had to wear Ace bandage wraps on my arms and a Jobst tailor-made compression garment for my legs to provide pressure and reduce the build-up of scar tissue. My plane travel was delayed until I could master getting the Ace bandages on my arms. With practice, I found I could do the task using my chin as an aid. To continue my physical therapy, I acquired an exercise bicycle and, within a couple of weeks, I was up to five miles a day. I also purchased a three-wheel bike to which I added a battery-powered radio. On the weekends, my son, Steve, and I rode around our neighborhood in El Lago. We would often end up with an entourage of bike riders.

A week after I was discharged from the hospital, I went back to work. Aaron asked me for position papers on a number of items over the ensuing year. The argument for manual control throughout the space shuttle flight regime was still being debated. W. H. Phillips, chief of the Flight Dynamic and Control Division at NASA Langley Center, discussed it at length in a white paper. Some of the merits of his memo strengthened my argument for an independent backup control system for the ALT. Another memo, from Milt Thompson at Flight Research Center, questioned the interest in a one-quarter scale orbiter test to assess subsonic aero. I felt that the test was worthwhile, and I was in favor of it if the estimated $200,000 cost was covered by OAST funds.

Andy Hobokan was the manager of the Manufacturing and Test Office, which was right down the hall from my office. His secretary was Patt Price. I didn't realize it at the time, but she would become my second wife.

One aspect of my job was overseeing the Engineering and Development Division's budget. In the Apollo days, with Max Faget in charge, the Apollo program's budget was between $70 and $90 million. During the space shuttle days, the budget was only about $20 million. In this role, I met all the E&D leadership in meetings and enjoyed working with the budget staff. We always scrambled to ensure that we would not lose any of our allotted funding at the end of the fiscal year. I instituted accounting on a cash-flow basis, rather than the old NACA process they were following. We reviewed the books monthly because we did not want any over expenditures to occur. To prevent cost over-runs, some projects were curtailed, or payments to vendors were pushed forward to the next fiscal year.

The Program Office continually dealt with unexpected circumstances that threatened the schedule. For example, Rockwell needed an immediate final layout for the space shuttle's instrument panel. The restriction of gravity on the crew's movement had to be considered. In order to solve the problem, I contacted a captain at the Naval Air Development Center in Pennsylvania that had a large operational centrifuge. I was relieved when he told me not to worry about the interagency paperwork approval and transfer of funds. A set of the various instrument panel segments were quickly fabricated and shipped. Joe Engle, Dick Truly, Bob Crippen, and Bo Bobko participated as subjects to determine reach capabilities at a 3.5 G level. Subsequently, Dick Truly led an evaluation of contingencies to determine what specific switches would be needed.

———

I consider myself to be an extremely fortunate man to have had a passion that tapped into my interests and abilities, and allowed me to draw from all of my life experiences. Changes were definitely in store for America's space exploration program. During this time, the last Apollo had flown, Skylab needed a shuttle to boost it to a higher orbit, and *Columbia*'s first shuttle mission to space was more than five years away.

———

CHAPTER 12

A RETURN TO FLIGHT TESTING

In 1976, I was one of the four astronauts who would fly the space shuttle orbiter *Enterprise* in the ALT program, which would take place at Edwards AFB. The JSC *Roundup* newspaper published an article about it, which prompted the fellow who supervised the center's parking spaces to pay me a visit. He sheepishly informed me that since I was slated to fly to space, it looked bad for me to park in a handicap zone. As a result, I lost my parking spot, which was better than the one that Chris Kraft, director of the JSC, had. Actually, it was time for me to once again pack up my desk and move back to Building 4 to share an office with Gordo Fullerton.

Those chosen for the two crews were Joe Engle, Dick Truly, Gordon Fullerton, and myself. Dick and I were in different crews, so I like to mention in presentations that *Enterprise* never flew without a Mississippian being aboard, since he also hailed from that great state. We had a superb test team forming under Deke Slayton and his deputy, Tom McElmurry, who had been chief of the Orbiter Atmospheric Test Office. I thought that this would be a good opportunity to include NASA KSC personnel who normally never worked directly on the spacecraft until it was delivered to KSC. I asked Charlie Mars to arrange for me to brief those who expressed interest in joining the ALT program. When I created the slideshow for the meeting,

Gordo Fullerton, me, Joe Engle, and Dick Truly in front of the *Enterprise*.
Courtesy of NASA

I cheated a bit. Since I knew the environment in the Antelope Valley high desert was considerably different from Florida, I only showed desert scenes in the early morning or near dusk, showing Joshua trees or beautiful fields of colorful wildflowers. I included great photos of Santa Monica Beach and the wooded picnic areas in nearby Tehachapi and the San Jacinto Mountains just east of Los Angeles. Ultimately, seventy-eight volunteered to transfer to California for the ALT program. The group took over the Butterfly Apartments in Lancaster that they called the KSC ghetto.

Rockwell assigned their best for test operations, many of whom I knew from the Apollo days. Rockwell's engineering and manufacturing expertise took place at their plant just down the road at Downey in Los Angeles. In Houston, a mini Mission Control was set up in what was the recovery operations room during the Apollo program. Don Puddy, with Harold Draughton, served as flight director for all the test flights. Others on the team were Dave Ballard, my friend from the Grumman LM days who was responsible for the flight plan, and Chuck Horstman. Guidance, navigation, and control were covered by Don Bourque, Harry Clancy, Herchel Perkins, John Nelson, L. D. Cheathum, and Ed Liebman. Chuck Deiterich, Mike Collins, and Travis Price covered flight dynamics and aero.

At Houston, during our first flight-crew get together, we assessed the situation we faced. There was a draft training plan, but no operational simulators. No procedures—normal or contingency—or mission rules existed. *Enterprise* was still being built and software was being developed. There was no central location for all the work underway that we had to cover. Engineering simulations went on at several facilities. Various activities took place from Houston to the west coast. Flight techniques, mission rules, procedures meetings, simulators/mockups, and the Shuttle Avionics Integration Laboratory were in Houston. Shuttle aircraft operations training took place at El Paso/White Sands and Edwards AFB. *Enterprise* manufacturing and testing was held at Palmdale Plant 42 and Edwards AFB in California. The NASA Ames Research Center Motion base simulator was at NAS Moffett Field and was our primary tool to refine the backup control system. The Engineering and the Avionics Development Laboratory were at Rockwell in Downey, California. It was apparent that our small group could not "cover the waterfront" if we tried to do it in unison. So, each of us was responsible for an area, but we brought in others as necessary. Communication between all members of various teams was key, so we wrote crew notes.

Joe Engle, the best natural pilot that I have had the pleasure to fly with, oversaw flight operations planning and the flight techniques meetings. He was also in charge of making sure that our primary simulators were ready. Joe flew development flights in the Gulfstream II shuttle training aircraft being developed by Grumman. Joe also worked the shakedown of the moving base orbiter aeroflight simulator, which replicated the *Enterprise* cockpit geometry and the control panel. A significant difference from our previous mission simulators was that it was equipped with flight-like computers. This enabled us to get a quick look at each new software release, particularly relative to the crew-interface aspects.

As it turned out, the toughest job of overseeing software development was led by Dick Truly. He was joined by Bob Crippen. They also had a lot of assistance from Ken Mattingly and Hank Hartsfield. I took on the vehicle readiness and test operations at the Palmdale facility, which was an extension of my previous Apollo experience. The first thing was to get things set up for our T-38 operations there. I carried out several sets of dust covers for engine intake and exhaust, tie downs, and canopy covers. I was happy to be greeted by Guenter Wendt, the "Pad Fuhrer," whom I worked with on close-out crews and who was there for Apollo 13's launch at KSC. He was in charge of on-the-floor operations at Site 1.

My first visit to *Enterprise* was on March 13, 1976. The cockpit was still being assembled, and only one of the computer screens and two keyboards were installed. Three of the four payload bay doors had been installed and fit-checked. Phil Shaffer informed us that the software required exceeded the memory available in the *Enterprise*'s IBM-101 computer. He reported that the OPS 1 prelaunch-module capacity was 60,960 words and free-flight OPS 2 was 64,509 words. Dick Truly stayed with the software issue through completion.

Joe attended the flight techniques meeting where they discussed the carrier's climb profile, chase aircraft requirements, *Enterprise* flight profiles, flight control ground rules, landing gear deployment, and strategies to handle the ground effect. The latter is something that cannot be accurately determined from wind-tunnel tests. The effect is encountered when an aircraft is within about one wing length of the ground. Possible variations include a vacuum sweep that could cause a hard landing or a "balloon" that pushes the craft away from the ground while running out of airspeed.

All flying machines, like people, have peculiarities. One example is how the F-84 cockpit floor vibrates when the underslung wing fuel tanks are depleted. Some have peculiarities with trimming, likely from being overstressed to some degree in their lifetime. The *Enterprise* was no exception when it came to unique personality traits. I wrote in my crew notes:

> After the Power Drive Unit for the vertical fin was installed,
> I observed the first motion of the rudder. It sounded like it was
> chewing itself up internally, even though the drive at the output
> shaft looked smooth. The Sundstrand company engineer present

said the sound was normal. The CRTs make a little sound like a little gear train in a windup toy. One hears a high-pitched whine when activating the Avionics bay and cabin fans until they get up to speed, and during one cockpit stint I got to cycle on and off the IMU fans, located about three feet forward and slightly left of the CDR's left rudder pedal. Startup noise is a noticeable screeching scream like the Avionics bay fans settling out to a steady roar.

There was an open house for all personnel, team members, and their families. I gave six talks to 670 attendees. Joe Engle spoke at the second open house to 1,000 attendees. We felt it was important for families to understand the space shuttle mission because of their sacrifices at home, and considering the ongoing twenty-four-hours-a-day, three-shift operation that we would experience in the weeks ahead.

I enjoyed seeing the *Enterprise* grow. In my crew notes from that period, I reported that there were three computer screens in place, as well as a forward replica RCS module. Many other systems were coming along as well, and the *Enterprise*'s raw gray-green metal had been painted black and white.

The closer we got to our launch deadlines, the busier we became. Joe and I used the T-38 and the shuttle-training aircraft to determine a minimum separation altitude. Through simulator practice, we developed a procedure to avoid exceeding the nose landing gear slap down limit of 10.5 frames per second if we had to use the direct control mode. We discovered that it took both crews to handle the control stick, pitch trim, and speed brake. Small panics arose such as the concern that a lack of venting in the nosewheel well during the steep approach, with the tail cone off, would cause a negative pressure, which would delay the timely deployment of the landing gear. The remedy for this was to add a piston cartridge attached directly to the nose gear strut to boost deployment.

The Rollout Ceremony took place on September 17, 1976. It was attended by the NASA administrator and some of the cast from *Star Trek*. There were short talks by the NASA administrator, a California senator, and Rockwell management, followed by photographs taken with *Enterprise* in the background.

Following all of the show biz brouhaha, we climbed aboard the *Enterprise* to check for outdoor window glare and general visibility. On January 31, 1977, it was ready to fly and was moved to Edwards AFB.

In the meantime, the entire crew took part in evaluation runs in the flight simulator for advanced aircraft (FSAA) at NASA Ames Research Center to define gains for the backup control system. The overall goal of using the FSAA was to set the gains in our backup control system to provide more stability and achieve a Cooper Harper pilot rating of four or five. In the Cooper Harper rating system, a pilot rating of one is perfect and a ten is unflyable. The simulator worked on a track to provide motion in two directions: fore/aft and left/right. The unique left/right motion provided a side force felt with lateral or roll inputs, comparable to that in the orbiter. Deke Slayton also participated. He brought John Denver, who was a licensed private pilot, with him one day. John's dad had been a USAF B-58 pilot. John made a few runs and quickly adapted to making successful shuttle landings. I made 159 runs with parameter or environmental changes between runs. The varied conditions included speed brake and body flap positions, center of gravity movement, and vehicle control through both the CSS and direct control modes. The spec turbulence seemed worse than any thunderstorm I experienced.

In addition to the many things that still had to be worked out, the crew continued training as per the training plan. There was ground school, shuttle training aircraft sorties, T-38 flights, and integrated simulations with Mission Control in the COAS moving base simulator. A full-scale orbiter in JSC Building 9 provided rappelling training, exiting from the hatch. This was the means of escape when the spacecraft was mated to the 747. Of course, we all gained firsthand systems knowledge through our participation in test operations as crewmen in the *Enterprise* at Palmdale.

A continuing problem was the multicomputer operation. The primary system was composed of four IBM computers that were devised to operate in unison, but the four computers could not operate together for very long. There was some discussion about whether multicomputer operations should be abandoned for a single computer to meet the flight schedule. The hope was the delivery of the IBM V23 software. Its release allowed the multi-computer operation to be rock solid.

Approaching flight time, a number of issues came out of the woodwork. One concerned the landing gear brakes. It was thought that a severe pressure spike could occur when the brakes were applied for the first time. The testing of the brakes and nosewheel steering had to be approached differently from that of a powered aircraft. An aircraft with an engine has the option of incrementally controlling the speed on taxi tests down a long runway. However, we were planning to perform the landing gear tests starting from the high-speed end after landing. Basically, the approach to the eight-mile Rogers lake bed runway allowed delayed use of the brakes and nosewheel steering on each landing. Gordo had assessed the *Enterprise*'s rudder-pedal brake application forces, and found that the left and right brake pedal forces for both crewmen had different force gradients. I called the project manager at Goodyear to find out whether they had a configuration of the cockpit brake pedals for us to try out. Unfortunately, their setup to qualify the brake system was performed solely under computer control. A similar brake-pedal force gradient situation existed in the COAS simulator, but a few runs in there convinced us that the different gradients didn't matter. The pilot could control the craft through the visual view out the window, so pedal force did not matter.

A 747 had been readied for the *Enterprise* tests. Owen Morris's Integration Office had conducted an early study of which specific model would be suitable. The choice was narrowed down to the Boeing 747 or the military's Lockheed C-5A transport. Ultimately, the 747 was chosen primarily because NASA could own the aircraft, which was not the case with the C-5A. Since it was a military plane, there was some concern that a national emergency would limit access to the C-5A. American Airlines put NASA in front of the line to meet the schedule. The aircraft with tail number N905NA became NASA 905. Boeing made a number of modifications to facilitate the *Enterprise*, including adding two tail fins to enhance directional stability. This was particularly important because of the possible failure of the 747 aircraft stability augmentation system. An escape tunnel that could be pyrotechnically deployed was added from the flight deck to a hatch at the bottom of the fuselage.

There was some concern about launching the test vehicle from the top of the carrier, because all of the prior experimental aircraft had been launched from below. Owen and the engineer, John Kiker, were both model airplane enthusiasts. They built and flew 1/40 scale models of the mated 747 and orbiter to prove that the separation concept was feasible.

The primary crew for the carrier aircraft testing was Fitz Fulton, commander; Thomas McMurtry, copilot; and Lou Guidry, Vic Horton, William Young, and Vincent Alvarez as flight engineers. Fitz was known as the commander of the various carrier aircraft for the X-series aircraft, from the X-1 through the X-15 and lifting body vehicles. I had worked with Vic Horton on several programs during my days as a NASA research pilot at Dryden Flight Research Center. Vic was the safety observer when I was flying our C-47A tow aircraft for Jerry Gentry's M2F1 lifting body checkout. Jerry was an Air Force pilot slated to fly the M2F2 that would be carried aloft by the B-52. Vic, as safety observer, made the tough decision to release Jerry when he got into a growing, roll oscillation on the tow line. Jerry managed a safe recovery to land without damaging the M2F1.

The 747 carrier aircraft tests got underway on February 15, 1977, with three taxi runs increasing in speed from eighty-nine miles per hour to 157 miles per hour on the Edwards runway. Fitz made the first takeoff of the combined vehicles on February 18, achieving an airspeed of 287 miles per hour and 16,000-feet altitude. Four more flights were made to evaluate handling qualities and performance with the last captive inert flight progressing up to 474 miles per hour and 30,000 feet altitude. *Enterprise* was unpowered for all the tests.

On June 18, 1977, Gordo and I were awakened at the Edwards Bachelor Officer Quarters for CA-1 or Captive Active-1, our first flight on top of the 747 carrier aircraft. A computer malfunction in preflight checks had caused a reschedule of the flight that had been scheduled the day before.

That morning, we had a quick preflight medical check and breakfast. Then we headed out to *Enterprise* where the test conductor advised us that the vehicle was only ready for the CDR ingress. I climbed into the left ejection seat, joining Dick Truly who had completed the preflight switch list. Dick reported that all had gone well and stayed to complete the computer reconfiguration and transition into OPS 2 mode. Gordo climbed aboard to join me an hour and thirty-five minutes before takeoff. Just before the tug push out of the 747 shuttle carrier aircraft (SCA) from the Mate-Demate Device, I could see a ground technician covering his nose with a handkerchief because he smelled ammonia. Our payload bay housed several large racks of ammonia bottles to support vehicle cooling, and I assumed and hoped that the odor was residue from servicing. As we backed out, I was impressed by how high up we were and, surprisingly, we could not see any part of the 747. The vehicle moved smoothly except for a barely perceptible square-747-tire effect (high-pressure tires going over taxiway joints). I felt the vibration of the engines starting.

As Fitz taxied out, the taxiway seemed very narrow from our vantage point in *Enterprise*. As we taxied, we encountered a number of alarms triggered by system-management software. I zeroed the faults shown on the screen just before takeoff. There were twelve alarms listed. For example, the air data redundancy management (RM) alarms came from a temperature differential from one air-data probe being impacted by the rising sun on one side of the vehicle. The tactical air navigation (TACAN) RM alarms were simply caused by the carrier aircraft changing directional orientation to cause blanking of various TACAN stations. I knew that we could work around these for ALT, but there would have to be some work on the system-management software for *Columbia* and orbital flight.

Fitz added the power for takeoff. Our runway acceleration was slower than I expected, but as the velocity increased so did our cockpit motion, particularly the side-to-side forces. We were at 140 knots airspeed when Fitz rotated to 17-degree pitch attitude. During the climb to 15,000-feet altitude, we heard a background low-frequency roar from the mated aerodynamics that continued throughout the entire flight, caused by the airflow between and around the mated vehicles. The auxiliary power unit (APU) whine, heard on the ground, was now masked by the aerodynamic airframe noise.

Nineteen minutes after takeoff, a flutter test was conducted. The orbiter controls had severe limits while mated to the SCA because of a concern that large movements would approach structural limits on the attachment structure. The pitch motion of the elevons was limited to 2.5 degree up and 1.5 degree down. Elevon movement in roll was limited to 0.5 degree motion. The rudder was also limited to 0.5 degree motion. Through all the contact "raps" that I did, there was no physiological response. Conversely, when the similar inputs were made by the SCA crew, we felt a surprisingly large lateral acceleration.

The pitch input response was small and the rudder insignificant with instant damping. This flutter test proved there was no structural instability and possible damage that could occur.

We deployed the speed brakes at twenty-three minutes after takeoff to address the possibility of severe buffeting and vertical tail failure, which was mentioned in a McDonnell study. Buffeting occurred at a 25 percent opening and it remained light up to 60 percent deployment. There was no buffeting increase when the speed brake was fully deployed and there was no vertical tail malfunction.

On the descent to landing, the CapCom requested that the RAM AIR valve be placed in the opened cabin pressure position. We heard a loud "whoosh" of air that caused the checklist pages to flutter. Surprisingly, we did not detect when Fitz dropped the landing gear and adjusted the flaps for landing, which was very smooth. We exited the 747 with the aid of a portable crane with a snorkel basket.

Joe Engle and Dick Truly's CA-2 flight plan called for more extensive flutter tests with the speed brake being incrementally deployed in and out through 10 percent increments. Fitz set up the conditions for a release to collect structural load data at the *Enterprise* attachments. The 747 also made a fly-by on the approach to the lake bed runway on a six-degree angle to test the microwave scanning beam landing station landing aid to support auto landing. Postflight inspection revealed a failure of the APU pump bellows seal with a loss of hydrazine fluid.

The third captive-active flight was flown as a dress rehearsal on July 26, with the primary objective of having the 747 fly the exact profile to set up the conditions for *Enterprise* separation. At that point, at 25,260 feet, I leaned over and could see the lake bed runway 17 that would be used for my first landing on the next flight. A couple of flight test objectives were accomplished in stowing and redeploying the air data probes along with deploying *Enterprise*'s landing gear on the 747 during the runway landing rollout.

Postflight, the nose landing gear spring failed, but had no effect on the deployment. This triggered a discussion on what the best actions would be should a landing gear not deploy. Testing this possible situation, I made several runs in the COAS simulator to discover that I could keep the wing from falling to the ground, using aileron and rudder control, until about 140 knots. After seeing this data, Joe and Dick decided that ejecting would be the best course of action. Gordo and I disagreed. I was not concerned about the ejection seat being adequate. My concern was the fact that ejection would necessitate pyrotechnics having to cut a hole through the vehicle's metallic ceiling to allow our ejection seats to clear.

As the date for free flight 1 approached, Rick Nygren, a NASA system engineer, distributed pertinent interim discrepancy reports, material review items, and unexplained anomalies. We referred to the latter as the funnies. They described unexplained readings or alarms that occurred in testing that could not be replicated with retesting. All of the appropriate subsystem managers assessed the results and discussed what could happen if various alarms occurred in flight. This gave us confidence before the flight-readiness review, chaired by Deke Slayton, that no lurking open items would hold things up.

Finally, the big day had come on August 12, 1977, to release the *Enterprise* from the carrier 747 to see how she flew. This was free flight 1. Five years had passed since I participated in vetting the proposal evaluations and selecting Rockwell to build the space shuttle. A new president, Jimmy Carter, had taken office in January, and we were a little nervous about his position on NASA, or his support of aerospace in general, because he had just canceled production of the B-1 bomber. Adding to these concerns was the fact that NASA had announced a major delay in the first orbital launch.

I felt a great weight on my shoulders that morning when I boarded *Enterprise*. I was not worried about the physical risk for Gordo and myself. The crew in the 747 were more at risk because any collision after separation might make it impossible for them to escape. Gordo and I had ejection seats. I was also focused on how the overall space shuttle program would be affected if we damaged or lost the *Enterprise*.

On ascent, Fitz and I carried out our various responsibilities. Fitz followed the oval racetrack profile to the north, while climbing to launch altitude. We conducted a flight control system checkout and the planned TACAN long-range test. Just prior to pushover, I noted an altimeter reading of 28,400 feet. The ambient noise level increased as Fitz gained speed in his slight descent. Gordo configured the wide-band recorder and cabin cameras. I set the solar electric propulsion, or SEP, pyro arm on and verified the SEP A and B lights coming on. I preset a commanded two degrees per second nose up pitch rate with my hand control to assist the expected upward movement and separation. Fitz gave the launch-ready call and I pushed the SEP button a moment later. We felt a sharp thump and acceleration of one G, but heard no noise from the pyros firing. Quicker than I would have thought, the chase aircraft called clear on vertical separation. As planned, I rolled right to a 30-degree back angle for horizontal separation from the 747, with a quick second chase aircraft call of horizontal separation. I was immediately impressed with the flight control response that seemed better than any we had experienced in our various simulation devices.

Initially, the four synchronized computers worked perfectly, but this elation came with another one of those *never panic early* moments, because we got a master alarm, a computer caution and warning, a CAM column on the GPC-2 computer, and a big X on CRT-2. Lo and behold, *Enterprise* was flying fine on the three remaining computers. Gordo was kept busy with the malfunction procedures to kill the blackballed Computer 2 by moving the GPS-2 computer mode switch to HALT and pulling circuit breakers.

After receiving the clearing calls, I pushed over to gain speed and estab-lish 250 knots to execute a practice flare leveling off at 20,000 feet. Chuck Dietrich in Mission Control had timed the speed decay to predict our lift-to-drag as being slightly low. Reaching an indicated nose boom of 11 degrees angle of attack and 185 knots, I lowered the nose to accelerate and transferred control to Gordo. I delayed taking over control to set up an overshoot to final that allowed a steep turn back to kill off some energy and set the speed brakes to 50 percent. Airspeed was increased in our shallow glide to stabilize at 280 knots. I retracted the speed brakes at 2,000 feet altitude with no apparent trim change. Preflare was started at 900 feet and Gordo armed and deployed the landing gear while passing through 265 knots to hear a muffled thump, with Bob Overmeyer in Chase 2 calling out that the gear were down. There was no notable ground effect perceived until we were at about 10-feet in alti-tude with a slight cushioning, leveling us off to coast just off the lake bed. Without further control inputs, *Enterprise* settled onto the runway at 185 knots. I fully opened the speed brakes and made a small pitch input to start the

I released the shuttle *Enterprise* from the back of the 747 for its first free flight on August 12, 1977. Courtesy of NASA

derotation. The orbiter is the only flying vehicle in my experience that one has to continue controlling carefully through the nose wheel on the ground to preclude a hard slap down.

As planned, only light braking was used during the 11,000-foot rollout. I made a few left and right rudder inputs at about 130 knots, and both Gordo and I tested the nose wheel starting at 100 knots and below. We could make heading changes without overshoot and could track well. This was contrary to what we had experienced in our OAS simulator, where the nosewheel steering control was loose with overshoots in changing direction. During the debriefing at the Dryden Flight Research Center conference room, the flight-summary data noted the total flight was five minutes and twenty-one seconds. Gordo's humor came through when he quipped, "This is not a program to build up your flight time."

Free flights 2 and 3 had similar profiles, with different ballasting for forward and aft centers of gravity. Each flight included a two-G wind-up turn to the final approach, reducing speed from 290 to 200 knots. Joe Engle and Dick Truly made the first tail cone off approach and landing on free flight 4. Free flight 5 had the major objective of achieving a precise landing on the Edwards AFB runway. A stripe was painted across the runway to mark the touchdown point. The 747 followed the oval racetrack pattern to the north and used the special rated thrust modification to achieve a release point at 17,000 feet and 245 knots airspeed for the straight-in approach to the runway. With vertical clearance, I started the right roll, and then turned back to align with the runway. I handed over control to Gordo, who pushed over to build up airspeed and deployed speed brakes corresponding to the microwave scanning beam landing system commands, achieving the desired glide slope at 9,600 feet altitude. I took control at 7,000 feet altitude and noted that we had drifted from our aim point on Lancaster Boulevard. To prevent going long, I steepened the descent and increased the speed brakes momentarily to 80 percent to prevent over speed. I selected a 55 percent speed brake approaching the 4,000-feet pull up and left them deployed because we were still high and fast approaching the runway.

The *Enterprise* speed seemed to hang up, gliding along at about four feet in the air as we passed the touchdown marker. I made pitch inputs to affect a landing 1,000 feet long at 180 knots airspeed, but *Enterprise* skipped back into the air, slightly rolling to the right. I tried to correct the roll but there was no response. I did not know at the time that my pitch inputs had caused rate limiting. Therefore, roll control was temporarily unavailable. I made a second and third attempt to correct the roll. The rate limiting was overcome, resulting in a large roll to the left. I went through several roll inputs over the next four seconds, out of sync, in what is known as a pilot-induced oscillation. At this point, Gordo suggested that I stop the control inputs, which I did, with the vehicle about five or six feet off the runway and airspeed bleeding off. I made a small pitch control input to achieve a second landing, harder than on any prior flight at four feet per second. During the remainder of the roll out, we applied moderate to heavy braking between 100 and 40 knots. The fix to the antiskid system was successful as no chattering was encountered. Needless to say, I felt embarrassed about this landing bounce and abnormal roll activity, particularly with a cast of several thousand people observing.

At a postflight party in Lancaster, I was awarded a small model of the orbiter, labeled *Enterprise*, with half of its right wing missing. After the program, a second gathering was held around the pool at the KSC Ghetto. I was afraid that the wives might throw me in the pool for luring their husbands away from home. Speaking of home, my marriage had been troubled throughout 1977. Many of the issues causing Mary and me to drift apart arose from the fact that, over the past ten years, I had hardly been home to spend time with my family. Mary had gotten tired of being both mother and father to the children, as well as taking care of all of the household matters. I couldn't even have told you how much money was in the bank. We mutually decided on parting and met with our children to tell them. I moved out in November. Our divorce became final in 1978.

—

Five months after the end of the ALT program, a press conference was held to announce the crews assigned for the upcoming orbital flights of *Columbia*. John Young and Bob Crippen were named to fly the first flight. Joe Engle and Dick Truly would fly STS-2, Jack Lousma and I were named to fly the third flight of STS-3, and Ken Mattingly and Hank Hartsfield STS-4.

We spent some time in the orbiter full-scale mockup. And once again, it was back to ground school to cover the systems in the *Columbia* that were not in the *Enterprise*. Over the next several months, we had classes that covered the environmental system, the propulsion systems, and software. On February 28, 1978, Jack and I had our first meeting about our assigned mission to rescue Skylab. Its orbit was decaying, and the vehicle was deserted after three crewed missions. Our STS-3 mission had the goal of executing a rendezvous: Jack remotely controlled the teleoperator retrieval system—a small booster that would be carried in our cargo bay—to fly over for docking with Skylab. Mission Control would determine the Skylab's orientation and fire the booster to either boost the lab to a higher orbit or deorbit it for a splash down in the ocean. To prepare for this mission, Jack and I made several trips to the Lockheed Martin Plant in Colorado to participate in the design and manufacturing of the booster.

Jack and I took advantage of any open time on the simulator, and I had a sprinkling of public affairs events left over from ALT. One day, Joe, Dick, Gordo, and I were all asked to join Chris Kraft in his office for a surprise. On February 17, we were given the NASA Special Achievement Award, which included $2,500. Later in the year, we all attended events in Washington and

California to receive the General Thomas D. White USAF Space Award, the USAF David C. Schilling Award, and the Society of Experimental Test Pilots Ivan C. Kincheloe Test Pilot of the Year Award. Advice from Eddie Rickenbacker prevented my head from swelling. He and I sat at the head table while I received the Gold Medal of New York Award in 1970 after Apollo 13. He told me that he had a basket of medals and said, "Never take yourself too seriously!" I understood what he meant, and I also knew that our success was a result of the sacrifice and hard work by many who worked on the program.

——

I married Frances Patt Price on January 9, 1979. I met her in the Orbiter Project Office in 1973. She preferred to use her middle name Patt—her mother wanted her name to be a little different, so she spelled it Patt with two T's. Patt had a son, Danny, and two daughters, Darla and Angela. Angela was nine years old and still living at home. Patt had grown up in the small town of Rogers, Texas. It had one traffic light, a post office, a bank, a gas station, a small grocery store, and a few cafés. Patt had an older brother, Bruce. Her dad did some farming and drove a truck to earn money to make ends meet. Patt and I took off from work to get married at the downtown Houston Municipal Court. Angela served as one witness and a court employee served as the second. From there, we went to lunch at the Spaghetti Warehouse, and, at Angela's request, went to see *Superman* at a Clear Lake movie theater. This was not a grand wedding ceremony and honeymoon, but it was certainly an unusual and memorable one. We both went to work the next day. Soon after, we moved into a two-bedroom townhouse with a loft. When my son Thomas visited for the weekend, the loft was his.

——

By 1979, it was apparent that the *Columbia's* first flight was not going to happen as scheduled. There were just too many problems. In the first ferry flight test at Palmdale to transport *Columbia*, a significant number of tiles came off. *Columbia* was delivered to KSC on March 24, 1979, and rolled into the Orbiter Processing Facility. There was so much unfinished work to be done that, for the first time ever, KSC signed the vehicle back over to JSC. The vehicle spent

610 days in the OPF. Then, the clincher that resulted in the Skylab rescue mission being canceled was the fact that its orbit was decaying faster than had been predicted. Skylab reentered on July 11, 1979. Some of its parts fell into a desolate area 300 miles east of Perth, Australia.

At that point, the way I looked at it, it would be at least two years before I would likely fly again, so I set my sights to move into executive management. Within NASA that path seemed limited for astronauts. Only Jim McDivitt and Jim Lovell had ever held management positions there. A few others who had left NASA before me had taken positions in a variety of fields—some far afield from airplanes and space with mixed success. I wanted to stay close to my passion and experience in aerospace, so I called Pete Conrad, who had joined McDonnell Douglas, to get a lay of the land. With his encouragement, I followed up with a call to George Skurla, whom I knew well from when he was in charge of Grumman LM Operations at KSC. George, at the time of my call, had risen to become president of Grumman Aerospace. I met with him in Florida at Patrick AFB, and he offered me the position of vice president of Space Programs at Grumman. I let him know that I was happy with the offer, but he asked me to wait until he talked to Chris Kraft and George Abbey, because he knew NASA well and didn't want to burn any bridges. Their discussion went well, and I submitted my resignation after twenty years with NASA. I filled out a Government Form 50, citing the reason for my departure as accepting "a promotion to managerial position in private industry." It was dated June 27, 1979, and it listed my parting salary as $45,792. I was sad to leave NASA, but the time had come.

CHAPTER 13

JOINING THE IRON WORKS

Patt, Angela, and I flew to New York and were met by a Grumman driver who took us to the Milleridge Inn in Jericho, Long Island, near the Grumman plant. We had Darla, Patt's other daughter, come up to visit, and she enjoyed lounging around the pool with the Blue Angels who were staying at the inn for their July 4 air show.

My family and I stayed in a suite until we found a home in the small community of Halesite in the town of Huntington on the North Shore. Patt was impressed by the small-town flavor of the community and Long Island in general. She was similar to many people who thought of the hustle and bustle of Manhattan at the mention of New York, but that's only a small aspect of the state. Our home was on Vineyard Avenue in a beautiful wooded area with no streetlights.

On my first day working for Grumman, I checked in at the personnel office on Oyster Bay Road to complete the customary new-employee paperwork. I also went to security to acquire a company badge. From there, I went to Plant 5 where I had spent many hours testing LMs. I met with George Skurla, whose office was at the end of what was known as Mahogany Row, so-called because of the hallway's beautiful wood. All of the executive offices were on the top floor of Plant 5 facing the runway. George and I reminisced about the days at KSC. He described Grumman's space business and shared his hopes for the future.

I also had a corner office facing the runway. It was the one Joe Gavin occupied when he headed the LM Program. From my Apollo experience, I knew many of the key executives and managers. Some had risen in the ranks. Joe Gavin was a vice chairman for Grumman. Aerospace was its primary business, but Grumman Corporation was still heavily involved in the aircraft production of six different Navy aircraft. Don Ingram gave me a quick rundown of the ongoing projects, followed by a tour of the first-floor areas, which contained a Sensitive Compartmented Information Facility, or SCIF for short, for classified activity.

Grumman's space business area was split into two sections—civil space and military space. The former included the manufacturing of wings for the space shuttle. One of my challenges was getting to know the people. All who worked in space business areas were design engineers, but none had worked in test operations during Apollo. The only one I knew was Dick Kline.

At each of the plants there were several reserved parking spaces for corporate officers. When I left work one day, my Volkswagen Beetle was missing. I called security and found out that transportation had towed my car away because the people there thought that no Grumman corporate officer would be driving a Beetle. A benefit of this embarrassment was that the Transportation Office thereafter took my car, once a week, for refueling and a wash.

In November, I accompanied Jack Bierwirth, Grumman chairman, and Pete Oram, vice president of Aircraft Programs, to Japan. Grumman had sold radar planes, called E2C AWACS, to Japan, and I was the appointed Knight on the White Horse. Part of my job was to portray a good image of Grumman at a number of events. There had been kickback scandals hanging over US aerospace companies for the sale of aircraft. My itinerary began at Grumman's Tokyo office in a building near the US Embassy. I gave several interviews to the press and was given a rundown of the activities that were planned.

The next day, I was driven to the Iruma Air Base for the opening day ceremonies of the Far East Air Show. Iruma is the home of the Japanese Air Defense Flight Group. Nearly 50,000 people attended the event. I spent most of the day in the Grumman Chalet, which was actually a large tent off the flight line. I enjoyed eating sushi, but quickly learned to stay away from the green hot sauce. In addition to meeting with the press, I also met with various individuals who were active in aeronautics and aerospace. One of them was the president of *Wings* magazine. He was an ace who flew the Mitsubishi A6M Zero in World War II. He also was shot down twice by the Grumman F6F Hellcat. The Grumman Hellcat and Vought Corsair were the first US aircraft that were competitive with the Zero in aerial combat.

The next day, I visited Gen. Takeda, chairman of the joint chiefs of staff, and Gen. Namatame, vice chief of staff, at the Japanese equivalent of the Pentagon in Washington. These were courtesy calls held in an impressive, immaculate office. Each day my schedule was full. From giving talks at places such as the University of Tokyo and receptions every evening, it was a whirlwind. I had dinner with an impressive group from the Keidanren, which was chaired by the head of Mitsubishi. The head of NASDA, Japan's space agency, was a member. This high-level business association had the mission of strengthening Japan's economy. I did not even have free time to be a tourist, but I was impressed with the cleanliness of the busy Tokyo streets.

Back home, most of my day-to-day activity was spent addressing new business. This was the most difficult part of my responsibilities, because in my career I was always involved in real-time operations or in a program underway. In growing space programs, the most important job was to help our government customer develop and promote a new program that would become a reality with government funding. Grumman, like all major aerospace companies, had an office in the Washington metro area, in Alexandria. Jack Herre was our man in Washington supporting space programs and keeping an eye on Congressional funding for programs of interest. There were a number of registered lobbyists who accompanied me in my meetings on the Hill or at NASA headquarters. Grumman's marketing department did a great job of identifying potential candidates to bid. The marketing department also kept an eye on our competitors.

As a result, there were many more business prospects available than the Space Program's annual funding allocation could support. Another major challenge was the constant budgetary shifting of priorities in the government. I found that half of the funds available each year went toward military programs.

Grumman and Boeing had a small contract to study a large Manhattan-sized array of solar cells placed in geosynchronous orbit, called the Solar Power Satellite. There were interesting challenges with the system. Microwave and infrared were being considered to transport the energy to Earth through relay stations in the ocean. That approach would not sit well with environmentalists, who would worry about how birds flying through the beam would be affected. It would also require the establishment of an air and sea danger area. In trade studies it was clear that construction would be better served by carrying up raw material for the beams, rather than prefabricated as was done for the International Space Station (ISS) construction. This concept, with marketing support, led to funding to develop a beam-building machine. When it was built, it was heavy and needed work to be flight worthy. It was brought to the MSFC for demonstration. It created lengthy beams that looked like Christmas tree ornaments, which hung from the ceiling in a large, high-bay building. Another challenge for the Solar Power Satellite to ever become a reality was the need for a booster-lift capability. It was thought that a heavy-lift vehicle could be developed via the "Star Wars" or strategic defense initiative missile defense system that President Reagan promoted, but the Star Wars program faded away and so did our funding from NASA and the Department of Energy. Gen. Jim Abrahamson was in charge of the Star Wars program, and I would cross paths with him later when he joined NASA as associate administrator for the Space Shuttle Program.

Through Grumman Data Systems, we had a contract with the Air Force Space Division. It was called SAWS (satellite attack warning system). As implied, it would support the monitoring of threats to operational US satellites. I enjoyed the opportunity to visit this wonder of a complex, which had a number of buildings under 2,000 feet of granite, supported on springs as protection from an earthquake or explosion. Ultimately, the Aerospace Defense Command (ADCOM), wanting their own contractor, forced us out by placing a mission-impossible timing requirement to track multiple targets in a fixed-price contract amendment.

Al Nathan led a contract that delivered two manipulator foot restraints to NASA. The manipulator foot restraint is the object seen in many iconic space photos where an astronaut is perched on the end of the Canadian Remote Manipulator System (RMS) arm working on the launch or repair of a satellite, as with Space Telescope and the Solar Maximum mission spacecraft. The Large Amplitude Space Simulator (LASS) was developed to support the design. It allowed subjects to see the dynamic effect on the RMS arm while performing simulated work tasks. Astronauts Jerry Ross, Don Linn, Bill Lenoir, and Bruce McCandless visited for an exercise on the LASS simulator. I made the mistake of granting Bruce a visit to our Research Department, which was working on a voice-aided system to reduce pilot workload in the F-14 fighter. Bruce, after listening to the PhD's description, said that you only need one voice command. With his deep, loud bass voice, he shouted out the word, "KILL!" The Research Department was wary of any future visitors that I invited for a tour.

In 1980, JSC signed a study contract with Boeing under Clarke Covington for the Space Operations Center (SOC). This contract was extended in 1982 to study a cheaper, evolutionary approach for assembly on orbit. Those studies' efforts provided the development of analysis data that ultimately led to a Request for Proposals and the award of contracts to four companies for the development of space station Freedom.

In September of 1980, with the assistance of Grumman International, I signed an agreement to support French Arianespace in marketing the sale of their Ariane rocket. Bob Renee was made manager and set up in an office with a secretary fluent in French. I enjoyed the opportunity to tour Arianespace's large manufacturing and test facility in France. We completed the sale of five Ariane launches over two years, satisfying offset credits for the sale of E2C aircraft.

I had an interesting diversion serving on President Reagan's Transition Team for space under George Low, former NASA Apollo leader and acting NASA administrator. Our instructions were to consider NASA's budgetary requirements on a scale of high, medium, and low. The Apollo program had the highest costs, and, during that era, NASA's existing activities required midsize expenditures, while the initiative with the lowest costs only included expendable rockets. In assessing the cost information provided, it was obviously that NASA's overhead was growing. All of the facilities that were in place during the Apollo program remained. Our final report avoided specifics to be implemented with the wording, "The question of whether or not NASA needs all the field centers should be addressed as soon as the purpose of the aeronautics and space program is defined."

—

On September 18, 1981, George Skurla received a phone call from Dick Smith, NASA KSC director, asking him to join him and other invited aerospace contractors in a meeting about a support-services contract for the space shuttle. This call resulted in a major career change for me two years later with the creation of Grumman Technical Services Incorporated (GTSI). In preparing for the contract bid, I spent a year in a two-trailer complex provided by NASA to all prospective bidders to study the overall launch-prep operations that were underway by a dozen contractors. I enjoyed learning about the ongoing work in areas that I was never exposed to as an astronaut. I received invaluable help from Dick Barton, who had been at KSC for Grumman during Apollo. Dick was the greatest planner I ever met. During his Air Force career, he worked with the development of the Minuteman nuclear missile sites and on war plans in the Strategic Command Headquarters. We, ultimately, joined the Lockheed team with Thiokol Corporation. In 1983, we won the fifteen-year shuttle processing contract. That involved my moving to Florida.

GTSI handled the unions and the requirement for a unique employee benefit package. I was named president. Our responsibilities included the Launch Processing System (LPS) in several of the launch control center firing rooms, which collected and processed the data to support the space shuttle launch. We also performed instrumentation and calibration of ground systems and the maintenance and analysis for the launch pad lightening protection system.

I had a *never panic early* moment, when I realized that we were to be the first of our team to transition to support the STS-3 launch on March 22, 1982. The transition process was exciting. It was also challenging to keep the varying degrees of chaos within relatively normal bounds, while not skipping a beat on the ongoing daily activity. This involved the recruiting and hiring of nine hundred employees. Some of the workforce was angry because of new changes in their benefits. I spent many long days visiting workers in their work places explaining the changes and ensuring that they were focused on the mission of supporting the next launch. At the same time, we were first up to establish an LPS at the Slick Six, which is what we called the Vandenberg AFB Space Launch Complex SLC-6, to support space shuttle launches into polar orbits.

Another first was that I would have to deal with the International Brotherhood of Electrical Workers, which comprised part of the workforce. Grumman never had a union before, and this was of some concern. However, I ended up dealing with a number of different unions on NASA and defense contracts over the succeeding twelve years and found that both the workforce and union leadership were dedicated to performing the work and mission required. To keep us on the right track I hired Lew Calvin for labor relations. His previous job with Disney during the construction of Disney World involved negotiating with nearly every trade union in the country.

Under other duties as assigned, Lockheed asked if I would join a NASA group known as the Blue Ribbon Crane Committee, led by Don Nichols at NASA. The committee had a goal of surveying cranes involved in lifting space shuttle hardware because there had been a Pershing II accident where a solid rocket segment dropped during a lifting operation, resulting in a fire as well as the deaths of three US soldiers. A series of reviews were held to analyze the causes. A study indicated the possibility that a solid rocket booster segment could ignite if it was dropped while hoisting in the Vehicle Assembly Building or VAB. As a result, new safety rules were established to limit solid segments in the VAB main aisle.

Al York, who was the Lockheed leader at Vandenberg, suffered a medical problem and John Denson, the Lockheed deputy at KSC, moved to hold Al's position for a year. In the interim, John requested that I fill in for him as chairman of our contractor change board at KSC. An unusual thing under the shuttle processing contract was that $5 million a year was available for the contractor to spend toward improvements in the operation. Pan American had a small contract with Lockheed to perform a cost-benefit analysis on changes proposed. As an example, I approved the installation of a laser-alignment system for the crawler. Before this improvement, the crawler would, usually, jockey back and forth a few times to align the mobile launch platform on its launch-pad supports.

Dick Smith, KSC Director, asked me to chair the annual three-day Space Congress Conference, held in Cocoa Beach, Florida. This took time away from my normal job, but one does not turn down a center director's request. This was the twenty-second year of the event, so a master schedule was available to facilitate the planning process. This was my fifth hat to wear, in my first six months on the job. My most important contribution was lining up speakers for the luncheon events and the banquet. The highlight of the conference was getting Ray Bradbury, noted science fiction author, to be the banquet keynote speaker. I took him on a tour of space shuttle operations and was amazed when we were at the Orbiter Processing Facility, viewing the underside of the orbiter, to see that he had tears in his eyes: This was his first experience seeing a real spacefaring machine.

———

On a day that was unusually cold for Florida, I joined some of our administrative staff at our Grumman office building in Titusville to observe the twenty-fifth space shuttle launch of STS-51-L *Challenger*, on January 28, 1986. One could not see the vehicle until it reached several hundred feet, and we were all shocked to see, at seventy-three seconds, a flash. Several components of the vehicle separated, flying off in different directions. I knew instantly that it was an explosion.

Doug Sargent, the new Lockheed leader at KSC, asked me to organize a review to ascertain if anything in our ground operations caused the *Challenger* accident. The Safety Advisory Board that I had chaired from the start of our contract would oversee the review. Members of the group included Willis Hawkins and Ed Cortright, senior retired Lockheed executives. Ed had led the Apollo 13 Accident Board. Others joining were Owen Morris, retired NASA head of the Space Shuttle Integration Office; Jack Enders, head of the Flight Safety Foundation; and Grant Hedrick, Grumman senior vice president. I scheduled briefings by key managers, across all segments of the STS-51-L launch prep operations. We also conducted one hundred face-to-face interviews with workers from various parts of the operation. Nothing was found that pointed to a processing error as the cause of the accident.

The next big program for NASA was space station Freedom, which ultimately became the ISS. Freedom was gaining momentum and provided my next adventure. The hardware would be procured through four Work Package contracts that would be aligned with four different NASA Field Centers. Work Package 1, won by Boeing, would be managed by MSFC; Work Package 2 was won by McDonnell Douglas under JSC management; Work Package 3, won by General Electric, reported to Goddard Space Flight Center (GSFC); and Work Package 4 was won by Rockwell Rocketdyne and managed by Lewis Research Center. Dick Kline, who had run Civil Space for me, called to say that the space station study had also included the need for a program office in Washington with a strong program manager to provide the system engineering and broad expanse of integration across four US contractors as well as three international partners. Dick also recommended acquiring a contractor to provide system engineering and integration support and a possible procurement opportunity for Grumman.

While attending the annual Space Foundation conference in Colorado Springs with my wife Patt, we met Chris Kraft accidentally at a gas filling station. What a small world. This was an incredible meeting, because Chris and his wife, Betty Anne, were not attending the conference. They were just passing through on their way to the Grand Tetons and Yellowstone National Park. I mentioned that Grumman was thinking of bidding on the space station program-support contract. When he heard this, Chris went off on a rapid-fire discourse about how program management from Washington was not going to work. He said, "Haise, you are crazy if you make that bid." His advice came back to haunt me many times over the ensuing four years.

———

Dick continued his work to line up the key personnel to support a bid for the program-support contract. A key element of a winning bid was to present the right management talent. The prevailing opinion was that I should be the director of the program-support contract to enhance a win. I had reservations about leaving the GTSI subsidiary, because I had just set up an infrastructure for it to grow and was facing a move to Washington, DC. Therefore, wanting to do what was best for the corporation, I "saluted the flag" and signed up to win the contract. Our marketing people determined that the competition was TRW with Lockheed as a teammate, and that Grumman would have a difficult time winning. After all, Dr. Fletcher, the NASA administrator, was TRW badge number 18.

Dick and I set up operations on the second floor of Plant 25 with a lineup of drafting boards utilized in the LM development days. Dick and I thought the environment might motivate our proposal team, reminding us that this was the last opportunity for most of us to play a major role with NASA. Ford Aeronautics and Booz Allen, our teammates, helped to prepare the proposal and NASA oral presentations.

After anxiously waiting for five weeks, Tom Moser, the head of the NASA Reston Space Station Program Office, called me on July 2, 1987, to let me know that Grumman had been selected as the program-support contractor. There were two challenges that our team immediately faced. One was tying down an office facility. The second was to quickly hire additional staff beyond the thirty-five people we had committed. In addition, we had Program Integration Offices to establish and staff to hire for NASA's MSFC, JSC, GSFC, and KSC facilities. In my first meeting on September 4, with Ron Mercer, our contract technical monitor, I was told that we were limited to hiring 260 personnel, but, in a meeting ten days later, Ron requesting a staffing plan that would result in seven hundred personnel by January. Because of this, the activity level could be best described by an old saying, "trying to run while pulling your pants up." A major recruiting effort was quickly established that included job fairs and newspaper ads nationwide at locations likely to produce the talent needed, namely engineers. Within ten days, a job fair had garnered the resumes of nearly five hundred people, half of whom were worthy of interviews.

———

There was another *never panic early* moment when I got a call from Scott Chrysler at corporate, informing me that Jack Bierwirth, the corporate chairman, had invited me to join him for the introduction of Grumman stock on the Japanese Stock Exchange. I had so many projects going at the same time, so the last thing I wanted was to run off to Japan, but an invite from Jack was more than an invite. Two weeks later, Jack Herre called to tell me that Senator Proxmire had "zeroed" space station funding in subcommittee. This was subsequently turned around in the full committee, though the NASA space station funding request was reduced by $267 million. Shortly thereafter, Ron Mercer informed me that there was a hiring freeze on for NASA and PSC.

The contentious Congressional funding deliberations continued throughout my four years with space station Freedom and always resulted in NASA receiving less than requested. The *Washington Post* and other national and local periodicals routinely carried stories of the space station budget struggles. The media did not help in our recruiting efforts or our ability to retain personnel. Another fallout effect of the reduced funding was the need to restructure to prevent schedule slippage. I oversaw three restructurings during my four years. Further congressional micromanaging occurred in fiscal years 1988 and 1989, where funding was limited until certain dates within each of those years. The second funding limitation was ridiculous. It required awaiting the election results to see whether the incoming administration and congress were for or against the space station. This was a clear indication of the complete ignorance of how such a measure would impact a program's momentum and costs, as well as the morale of all personnel.

There was also continual turnover in NASA management while I was on the job, including three administrators at the top, three space station program managers at headquarters, and three space station directors in Reston, Virginia. I reported to the space station director that was, initially, Tom Moser. Tom was replaced by Ray Tanner from MSFC in January 1989, who, in turn, was replaced by Bob Moorehead in September 1989.

I met with Sy Rubenstein, who was in Reston for a visit with Bob Moorehead. Sy told me that the Grumman integration effort was going to the centers and that it was "impossible to work NASA in my knickers." Bob had expressed that he felt Kline and I were the wrong people to get the job done the way he wanted. An internal Grumman meeting was held, and executive leadership arrived at the consensus that a change in management was needed. Tom Kelly was chosen to replace me at Reston. Because of this chain of events, I knew the happiness that Brer Rabbit felt when he conned Brer Fox into throwing him back into the briar patch thicket. I felt bad about leaving Tom with the mission impossible, but was happy to get back to the GTSI briar patch in Florida. I have come to look at those years of working on space station Freedom as my time in purgatory, which is defined as "a place or state of temporary suffering and misery." I recalled from my Catholic upbringing that suffering was good for my soul.

Back at GTSI on May 1, 1991, Wiley Williams, who had replaced me four years prior, arranged a set of briefings to cover the status of current programs. Wiley briefed me on an adventure he had gotten into with a contract directly to Saudi Arabia through a prince. The opportunity had been set up by Grumman International, and he had been hesitant about having a contract directly with a foreign country. Grumman had a workforce of 450 personnel to support the operations and maintenance of all the ground radar system across Saudi Arabia to support both air traffic control and defense. Wiley was dealing with a recurring problem of slow payment of the payroll invoices submitted. Skip Olson, director of our Shuttle Processing contract, reported that things were going well. Under Skip's leadership, Grumman won the NASA 1991 George M. Low Award in recognition of quality and performance.

John Philips, GTSI marketing director, had crafted a winning bid for the US Navy's Advanced Training Command TA4J and navigator training aircraft. The contract covered seven hundred employees located at NAS Pensacola, NAS Meridian, NAS Chase Field, NAS Kingsville, and Naval Air Facility El Centro. I met with our program director, Tom Scott, a retired Navy captain, for a briefing on his futile efforts with the contracting office at CNATRA Headquarters in Corpus Christi. I followed up with tours at all the training sites to meet our site managers and Navy management. It was great to be around active flight lines, smelling the kerosene and hearing the shrill jet engines. One thing that hit me right away was seeing how young the pilots were striding out to their aircraft. In meeting the Navy management at NAS Kingsville, I noted that I had gone through flight training there thirty-eight years before and realized that I looked that young then. The base commander quipped that even his desk was the same. I mentioned that I had done most of my flying as an instructor at South Base. The commander told me that South Base was gone, but he had a driver take me to where it previously was to take a look around. Here and there

I could see piles of broken asphalt and concrete from what used to be runways and taxiways. We stopped to look around, and I nostalgically thought back to those days when I was roaring off in an F6F, T-28B, SNB, or S2F aircraft.

The next challenge under my leadership was winning the JSC Information Services Contract in November 1992. Our team included Boeing Information Services and the SAIC Corporation. The contract required 650 personnel to serve as support, 10,000 desktop/laptop computers, commercial off-the-shelf software, main frame computers, and a 24-hour help desk for the JSC. Bud McKenzie, who had been in a similar assignment for Grumman Corporate, led our team. This experience opened my eyes to the importance of PCs in the workplace.

During this era of my life, I was also involved in fact-finding meetings with Martin and Northrop personnel, because both companies had made offers to acquire Grumman. Since defense budgets were declining, companies such as Convair, North American Aviation, Douglas, Martin, Chance Vought, Fairchild, and McDonnell Douglas had disappeared. Ultimately, Northrop offered $62 a share. This acquisition resulted in Northrop Grumman.

Northrop had a long-standing service company, named NWASI, or Northrop Worldwide Aircraft Services Inc. After the merger, I continued as head of the GTSI subsidiary, and I also managed NWASI, which employed approximately 1,600 people, half of whom worked on the Vance AFB aircraft maintenance contract at Enid, Oklahoma. Another NWASI contract supported Army aviation elements, vehicle maintenance (motor pools), supply warehousing, and cooks at Ft. Eustis, Fort Monroe, and Fort Story in Virginia. Dick Cochran was the director, and our workforce supported deployments of five hundred Army personnel to Haiti after an earthquake and six hundred to Iraq.

When companies merge, the early challenge is to reduce overlapping personnel and systems, as well as to retain key people, while not impacting any of the ongoing operations and contracts. Our GTSI/NWASI challenge was having two headquarters for two small subsidiary companies. We made the easy decision to maintain the contracts that were in effect to avoid the hassle of novating government contracts. Personnel, though not many were impacted, was a tougher issue. When I realized that a recompete was scheduled for the major Vance AFB contract in a year, I made a number of visits to Vance to meet with Ron Shamblin, retired USAF colonel, who was the director for the contract. He toured me to all the workplaces and introduced me to Col. Soligan, who was the USAF commander at Vance. Again, it was great to be around the flight line and hangars with lines of T-1s, T-37s, and T-38s. Ron Bartlett, of business development, informed me that we had won the Vance competition for another five-year contract on July 5, 1995.

A combination of things led to my decision to retire after seventeen years with Northrop Grumman at the age of sixty-three in the summer of 1996. Primarily, it was apparent to me that Northrop, like Grumman leadership, did not have service business as a priority in their strategic plans. The heart and soul of both companies was in the winning of contracts to design, develop, and build new aircraft and electronics. As I look back over my experience in the business world, I would say that, due to the workload and stress, it was as complicated as my experiences as a pilot and astronaut. Secondly, with the diversity of skills and variety of people in the workforce, I was often involved with too much social engineering dealing with people problems.

CHAPTER 14

IN THE ROCKING CHAIR

My mom had always advised me not to just sit in my rocking chair after I retired. She followed that advice herself, still driving her great-grandchildren to and from school in her late eighties. My first venture back into aerospace was signing up as a consultant with Northrop Grumman to take part in the Red Team review of their proposal for the *Orion* contract. I read the executive summary, and it was my opinion that the prime contractor and integrator should have had the responsibility for the capsule and all the software. Unfortunately, the teaming agreement had already been signed and sealed. That experience ended my career as a consultant. I found it unsatisfactory to come in after the decisions had been made and to have no authority.

—

We bought a small home in Gautier, Mississippi, on Mary Walker Bayou, for some relaxation and good fishing, my one hobby. Patt called the home my fishing hangout. I got a surprise call from Leo Seal, head of the largest bank in south Mississippi, to talk about joining him on a newly formed not-for-profit's board. Leo shared with me some of the history that led to the project. NASA Stennis Space Center, about forty-five miles from Biloxi, had a small museum onsite, called StenniSphere. The 9/11 tragedy and increased security requirements for government facilities severely impacted StenniSphere's attendance. Leo told me that in his discussions with Roy Estes, the Stennis director, and Myron Webb, of Stennis's Public Affairs Department, they believed a facility was clearly needed offsite. As part of the plan, NASA would provide eighteen acres under a thirty-year land-use agreement with the Infinity Science Center's board. The challenge for the board was to raise $40 million to build the facility and the planned exhibits. I wasn't sure how much I wanted to be out of my rocking chair, but after reading the short write-up covering the project, I agreed to join the board of directors. Since I was born and raised in Biloxi, it struck me that my board membership was a community service contribution that I could make after being away from home chasing my career goals for forty-five years.

I saw several benefits of Infinity Science Center. First, it would be visible from Interstate 10, where more than eight million cars drove by each year, from Louisiana to Mississippi. Obviously, it was a nice icon for Mississippi's Gulf Coast tourist industry. Second, the upstairs Space Gallery that was planned would tell the NASA story and the work being conducted at Stennis Space Center to certify rocket engines for flight. The last benefit of the project was what sold me. Infinity would provide a fun, educational experience to visitors— children, in particular. The Infinity Science Center and other space and science museums offer learning experiences about the incredible achievements that arise from technology.

For more than a decade, it has been rewarding to be a part of Infinity's leadership. I proudly witnessed the opening in the spring of 2005 and have been impressed by the various STEM education programming such as Space Camp, Rocket Day, Boardwalk Nature Journals, and Exhibit Scavenger Hunts. STEM experiences for children should be top priority in our society, because in the 2019 Program for International Student Assessment Survey, the US ranked thirty-eighth in math and twenty-fourth in science. Clearly there needs to be an ever-increased effort to ensure that the United States remains a competitive global leader. Fortunately, NASA, with its exciting missions for exploration and its problem-solving to contend with Earth's global threats, will always attract the most talented people who gravitate to challenges.

I wonder what challenges my grandchildren and great-grandchildren will face in the century ahead. I envision major threats that space can help

Infinity Science Center, located just off of I-10 near the Mississippi-Louisiana
border, engages students and adults alike. Courtesy of Infinity Science Center

mitigate, either through satellite monitoring and data collection or providing
a direct solution such as altering the path of an asteroid. NASA has a page on
its website devoted to climate change. A casual perusal of the page leaves no
doubt that human activity is causing the Earth to warm. The site shows that
the warming trend has grown over the past two centuries, with the greatest
amount of the warming occurring in the past thirty-five years. Listing shrinking ice sheets, glacial retreat, sea-level rise, and extreme weather events, NASA
predicts that sea levels will rise by as much as one to eight feet by the year 2100.
NASA also predicts an ice-free Arctic in summers and more intense weather.
While there is extensive coverage in the media about the possible effects of
climate change, there are disagreements among experts and nonexperts on
how to control this mounting crisis. Some think that the primary cause is
the ever-growing human population.

Regarding the environment and climate change in relationship to the
burgeoning world population, the numbers are telling. In 1930, just before
I was born, the world population was 2 billion, growing to 4 billion in 1974,
6 billion in 1994, and 7.8 billion in 2020. The enlarged population results
in more carbon dioxide and greenhouse gases, due to increased uses of fossil

fuels in transportation and energy generation. Humanity's needs have led to the use of more fertilizer for higher crop yields to support increasing food requirements. A secondary environmental effect of fertilizer is the possible damage to water aquifers. The increasing population has required more living area, which leads to deforestation and, in some cases, has threatened or caused the actual extinction of some animals.

I have noted from history that, normally, governments of any type only act promptly and adequately during a crisis. I suspect that there would be worldwide action and a change in the *never panic early* philosophy, if climate change results in a foot of water in most coastal cities. I do not know what would be required to remedy such a catastrophic situation, but it would be a great project to manage. Because of its urgency, one could easily acquire a talented, motivated workforce to create solutions, which would be adequately supported by governments.

Another hidden threat that is not covered much by the media is the growing number of countries possessing or developing nuclear weapons. Most people do not give this reality much thought. I didn't for quite some time, but my call to active duty with the 164th Tactical Fighter Squadron during the Berlin Crisis in 1961 was a sobering experience.

Over the past few years, Col. Jack Anthony has arranged for me to speak to some of his students at the Air Force Academy. One class was involved in a project to develop CubeSats, labeled that way because they are small enough to ride on a rocket as a ride-along addition to the major satellite component. They rode piggy back on various USAF launches. I subsequently heard that one of the cadets—a young woman—after graduating, received her first assignment to a missile silo complex. USAF operations officers are on duty around the clock, seven days a week, at missile silo control stations across the United States. Fortunately, there is a safeguard to prevent the unauthorized launching of a missile. It takes the coordination of two people to actually launch one.

The US started the world down this dubious trail with the use of nuclear weapons on Hiroshima and Nagasaki in 1945. During that same period, Germany and Russia were developing nuclear weapons. Over the years, the number of countries with nuclear weapons has grown to include England, France, China, North Korea, Pakistan, India, and Israel, and more nations are working to create their own nuclear weapons. Six nuclear weapons were lost by the United States in the 1950s and '60s, and have never been recovered. One rolled off the deck of a ship; others have vanished in plane crashes. Under the Treaty for the Nonproliferation of Nuclear Weapons, the International Atomic Energy Agency is responsible for keeping track of the weapons and materials of the member nations.

Hollywood has made a number of movies about the threat of meteors over the years. Two relatively recent ones are *Deep Impact* and *Armageddon*. Many viewers of these films consider them to be mere science fiction, and are unaware that a meteor impacting the Earth is a disaster that can be avoided. My friend and fellow astronaut Rusty Schweickart, who flew the first LM on Apollo 9, has devoted many years to educating the government and the general public about this threat.

In 2002, Rusty, along with three-time space shuttle astronaut, Edward Lu, formed the B612 Foundation. Its mission is to protect Earth from asteroid impacts. A possible real impact by the asteroid 9942 *Apophis*, predicted to happen on April 13, 2004, got a lot of press attention. However, through refined tracking data, that threat was dispelled. With the hope of stirring up action for the development of an asteroid defense capability, Rusty met with a congressional committee in May 2005 on the subject.

In April 2018, the B612 Foundation reported, "It is 100 percent certain we'll be hit [by a devasting asteroid] but we're not 100 percent sure when." An example of actual impacts that many have heard or read about is the Chicxulub meteor impact near Yucatan in Mexico, which is cited as the cause of the extinction of dinosaurs and three quarters of the plant and animal species approximately sixty-six million years ago. In recent times, the Shoemaker-Levy comet was in transit through our solar system, but was trapped in orbit around Jupiter. We witnessed, firsthand, the Shoemaker-Levy comet breaking up and impacting Jupiter over six days in July 1994. The combined energy of these impacts was estimated to have the energy equivalent of six million megatons of TNT or six hundred times that of the entire world's nuclear arsenal. A system to defend against this possibility is certainly feasible with today's technology. A detection and tracking effort has led to the identification of an estimated 90 percent of all asteroids over one kilometer in size. In 2005,

Jim Lovell (*right*) and I reprise our end of mission wave at the unveiling of the Apollo 13 astronauts statue at JSC in Houston in 2021. Courtesy of Angela Alexander

Congress tasked NASA with detecting and cataloguing 90 percent of the estimated 25,000 meteoroids 140 meters and larger. Even if the Near-Earth Object Surveyor mission is funded to launch in 2026, it will take several years to complete the search. The unexpected appearance of asteroids and comets from interstellar space, with Oumuamua being sighted on October 19, 2017, passing through our solar system, is a low-probability threat, though it received press coverage.

Amazingly, a number of spacecraft have rendezvoused, orbited, and landed on asteroids and comets. Our Near Earth Asteroid Rendezvous mission resulted in an orbit of the spacecraft around the asteroid 433 Eros in February 2000. The NEAR Shoemaker spacecraft actually landed on the asteroid Eros. On the European Space Agency Rosetta Mission, photos were taken of the comet 67P/Churyumov-Gerasimenko from as close as sixty-five feet before landing on the surface. The most recent OSIRIS-Rex Mission collected a sample from the near Earth asteroid Bennu. A defense capability clearly calls for international cooperation, because any impact affects the entire planet. A mission to look at that capability is the planned DART or Double Asteroid Redirection Test. It is planned to rendezvous with the asteroid Didymos that has a smaller satellite asteroid in orbit around it. An impact on the small satellite by the small DART vehicle will allow the analysis of that approach to altering the satellite's path around Didymos.

We are blessed with talent that can overcome challenges like those we faced in the Apollo program. Hopefully, this talent will be focused globally to solve future challenges threatening our species. I also hope we continue our destiny for exploration of the dynamic universe that the Creator has provided. I just don't have a large enough ego to go out on a clear night, view all those dots of light representing millions of stars and galaxies, and think that the Creator put all that there just for me.

———

ACKNOWLEDGMENTS

A lot of people helped us as we worked on *Never Panic Early*. Early in the process, Bill Barry, NASA chief historian (now retired), provided some crucial advice on where to find the NASA records we needed. Staff members at the NASA headquarters History Office, including Elizabeth Suckow and Colin Fries, helped us in our search for hard-to-find information. We also turned to fellow astronauts Tom Stafford and Walt Cunningham to confer about different topics and the business of writing a book, and we thank them for their insight.

Also, early in the development of this book, literary agent Deborah Hofmann provided great direction. She gave us the confidence to be able to set course and sail forward with the story we had. In the same fashion, Rick Houston, fellow author and space historian, provided good advice on the publishing world. Rick's friend and ours, Milt Heflin, is someone we often turn to in order to find out details about missions or who's who in the space world. Milt was an Apollo spacecraft recovery specialist and transitioned to become a flight director before going on to bigger and better things at NASA. We also thank fellow author Francis French for passing along some words of encouragement at the right time. We truly appreciate the great flight director Gene Kranz for writing the foreword to this book. What a wonderful friend!

A special group at the University of Houston-Clear Lake's Archives and Special Collections oversees a marvelous archive of Johnson Space Center documents and history. The people who work there were a great help, and the documents we located were just what we needed. Thanks to Chloris Yue, director of Library Public Services, and archivists Tamatha Brumley and Chieko Hutchison. Similarly the archives folks, Charles Sullivan and Karen Taylor, at Mississippi Gulf Coast Community College were extremely helpful in finding photos and information on my time at Perkinston Junior College. At the Nixon Presidential Library and Museum, we received some great help with pictures and archives from Lina Ortega and Marcie Kissner, curators of the Western History Collection. The staff at the University of Oklahoma Libraries did a lot of work for one yearbook photo of me and class information from 1959, and I appreciate it.

George J. Marrett, author and fellow member of Aerospace Research Pilot School, Class 64A, consulted with me on the process of getting a book printed. Dick Kline and Connie Blyseth were invaluable for information about the Space Station Freedom system engineering and integration contract experience. John Phillips and Bob Rennie were extremely helpful with details on the Grumman marketing effort to sell Ariane rockets. Lori Morrison of Northrop Grumman Corporation and Pamela Griffin-Hansen, librarian archivist, provided photos of lunar modules in their test and manufacturing stages. Thanks to Bob Sieck who provided me with the list of Kennedy Space Center personnel who worked on the Approach and Landing Test program. Bob, as an *Enterprise* test conductor, also visited with me about the testing that went on in order to get *Enterprise* ready for flight. And thanks to Charlie Mars, who provided data on his participation in the meeting that led to Apollo 8 going to the Moon. Charlie also discussed the sad closure of space station Freedom and the fate of civil servants with me.

At Smithsonian Books, we had such professional and enthusiastic help and guidance. Our project manager and Smithsonian senior editor, Jaime Schwender, was there for us to see it through to the end, answering the tough questions and guiding the project. Our editor, Karen D. Taylor, provided the trained eye to help with adjustments to the book. Matt Litts and Sarah Fannon provided great professional guidance on the marketing aspects of the book.

Acknowledgments

Carolyn Gleason, director of Smithsonian Books, provided the leadership to make this project a reality. Assistant editor Julie Huggins provided great support for this book to be published and we truly appreciate her enthusiasm.

This book has been a joint effort. There is no way I could have done it without my coauthor, Bill Moore. I would like to extend a special thanks to our families in Texas, Oklahoma, and Mississippi for putting up with interruptions in our daily lives for phone calls, research time, trips, and writing. We've enjoyed the journey and look forward to sharing this book with all those around the world who love a good story.

INDEX

Index

Index

Index

Index